VERPOOL JMU LIBRARY
3 1111 01464 1599

Life Cycle Costing for the Analysis, Management and Maintenance of Civil Engineering Infrastructure

Edited by John W. Bull
Professor of Civil Engineering, Brunel University, UK

Whittles Publishing

Published by
Whittles Publishing,
Dunbeath,
Caithness KW6 6EG,
Scotland, UK
www.whittlespublishing.com

© 2015 J. W. Bull
ISBN 978-184995-148-7

All rights reserved.
No part of this publication may be reproduced, stored in a retrieval system, or transmitted, in any form or by any means, electronic, mechanical, recording or otherwise without prior permission of the publishers.

The publisher and authors have used their best efforts in preparing this book, but assume no responsibility for any injury and/or damage to persons or property from the use or implementation of any methods, instructions, ideas or materials contained within this book. All operations should be undertaken in accordance with existing legislation, recognized codes and standards and trade practice. Whilst the information and advice in this book is believed to be true and accurate at the time of going to press, the authors and publisher accept no legal responsibility or liability for errors or omissions that may have been made.

Printed by

Contents

Contributors

Professor Adisa Azapagic
School of Chemical Engineering and Analytical Science,
The University of Manchester, Manchester, UK

Dr Astrid Bjørgum
SINTEF Materials and Chemistry, Postbox 4760 Sluppen,
NO-7465 Trondheim, Norway

Ir Prof Moe M S Cheung
Western China Earthquake and Hazards Mitigation Research Centre,
Sichuan University, Chengdu, China

Dr Rosa M. Cuéllar-Franca
School of Chemical Engineering and Analytical Science,
The University of Manchester, Manchester, UK

Dr Helena Maria dos Santos Gervásio
ISISE, Department of Civil Engineering, University of Coimbra, Rua Luís Reis Santos,
Coimbra, 3030-788, Portugal

Professor John T. Harvey
University of California Pavement Research Center,
Department of Civil & Environmental Engineering,
University of California, Davis, USA

Matthias Hofmann
SINTEF Energy Research,
P. O. Box 4761 Sluppen,
No-7465 Trondheim, Norway

Dr Changmo Kim
University of California Pavement Research Center,
University of California, Davis, USA

Professor Eul-Bum (E.B.) Lee
Graduate School of Engineering Mastership,
Pohang University of Science and Technology, Korea

Andrew Moore
Life cycle Logic, Fremantle, Australia

Andrew Pascale
School of Engineering and Information Technology,
Murdoch University (South Street),
Western Australia, 6150, Australia

Associate Professor Filippo G. Praticò
Department of Civil Engineering, University *Mediterranea* at Reggio, Calabria, Italy

Professor Anand J. Puppala
Department of Civil Engineering, The University of Texas at Arlington, Texas, USA

Associate Professor Sireesh Saride
Department of Civil Engineering, Indian Institute of Technology, Hyderabad, India

Ir Dr Kevin K L So
Faculty of Science and Technology,
Technological and Higher Education Institute of Hong Kong

Dr Tania Urmee
School of Engineering and Information Technology,
Murdoch University (South Street),
Western Australia, 6150, Australia

Dr Thomas Welte
SINTEF Energy Research,
P. O. Box 4761 Sluppen,
No-7465 Trondheim, Norway

Chapter 1
Life cycle cost analysis of the UK housing stock

Rosa M. Cuéllar-Franca and Adisa Azapagic

1.1 Introduction

A number of studies have considered life cycle environmental impacts from the housing sector (e.g. Adalberth, 1997; Adalberth *et al.*, 2001; Peuportier, 2001; Asif *et al.*, 2007; Hacker *et al.*, 2008; Hammond and Jones, 2008; Bribián *et al.*, 2009; Oritz *et al*, 2009; Mohan and Powell, 2010; Cuéllar-Franca and Azapagic, 2012) but the life cycle costs have seldom been addressed. And yet, economic aspects such as housing costs and affordability are important for the sustainable development of the residential construction sector.

The housing sector is very important for the UK economy as it directly affects the economic growth (HC, 2008). For example, in 2010, the construction industry contributed 8.5% of the UK's total gross domestic product (GDP) of £1.45 trillion, to which the residential sector contributed 40% (UKCG, 2009). After Denmark and Greece, the UK has the highest housing prices across the European Union with people spending around 40% of their income on housing costs such as mortgage payments and energy bills (Eurostat, 2012). The latter is the cause of fuel poverty of around six million households owing to the rising energy prices (DECC, 2009; Bolton, 2010).

In recent years, many people have been unable to purchase a home because of changes in the availability and types of financial and mortgage products (Sergeant, 2011; DCLG, 2012; RICS, 2012). This situation has created an unstable housing market, which has led to a fall in house prices, and dragged the UK economy further into recession. For example, the average house price of around £190,000 in 2008 fell to £160,000 in 2011 (HPUK, 2012). Home ownership is also declining and in 2011 it dropped to 66% from 70.9% in 2003; so the proportion of households that own their own homes has fallen back to where it was in 1989 (BBC, 2012). This trend is expected to continue over the next 10 years (Sergeant, 2011). Such a situation is affecting particularly young people – only 10% of all owner occupiers are under 35 years of age (BBC, 2012) while 33% of first-time buyers are over 35 (DCLG, 2012).

It is therefore important to understand the full costs of housing and their main sources along the whole supply chain. This is the subject of this chapter, which sets out to estimate the life cycle costs of the current housing stock in the UK and identify cost reduction opportunities. Three typical types of houses are considered: detached, semi-detached and terraced houses (Utley and Shorrock, 2008). This work complements a previous study on life cycle environmental impacts of the current housing stock in the UK (Cuéllar-Franca and Azapagic, 2012), which are also briefly discussed later in the chapter as part of an improvement analysis. As far as the authors are aware, this is the first life cycle cost study of the housing sector in the UK. Elsewhere, only two life cycle costing (LCC) studies of individual houses have been found in literature, one based in Finland (Hasan *et al.*, 2008) and another in the USA (Keoleian *et al.*, 2000).

1.2 Methodology

As there is no detailed LCC methodology for the residential construction sector, the methodology used here represents a combination of the general guidelines available for the building sector such as the ISO 15686-5 (BS, 2008) and EN 15643-4 (BS, 2012) and the approaches described by Swarr *et al.* (2011), Gluch and Baumann (2004), Hasan *et al.* (2008), Rebitzer *et al.* (2003), Abeysundra *et al.* (2007) and Hunkeler *et al.* (2007). To a large extent, the LCC methodology adopted here is congruent with the life cycle assessment methodology (ISO, 2006a,b).

1.2.1 Goal and scope

The goal of the study is to assess life cycle costs of the current housing stock in the UK and identify opportunities for improvements. This is carried out by first estimating the life cycle costs of individual houses considering detached, semi-detached and terraced homes. These results are then extrapolated to the UK housing stock consisting of seven million each of semi-detached and terraced and four million of detached houses (Utley and Shorrock, 2008). Collectively, this represents 72% of over 25 million residencies with the rest being apartments in multi-storey buildings, consideration of which is outside the scope of this study.

As indicated in Figure 1.1, the system boundary for the study is from 'cradle to grave', including all activities from extraction and manufacture of construction materials to construction and operation of the house to its demolition. The functional unit is defined as the construction and occupation of a house in the UK over the lifetime of 50 years. The following typical usable floor areas are considered (Brinkley, 2008):

- detached house: 130 m^2,
- semi-detached house: 90 m^2, and
- terraced house: 60 m^2.

Like the large majority of UK houses (92%), they are assumed to be built in a traditional way, with strip footing foundations, brick external walls and pitched roofs with concrete tiles (Brinkley, 2008; DCLG, 2008). The average usable floor area across all three types of house is equal to 93 m^2 which compares well with the average 91 m^2 for all UK houses (DCLG, 2008). It is also assumed that each house is occupied by an average UK household consisting of 2.3 people (Utley and Shorrock, 2008). Therefore, the houses considered in this work are representative of the whole UK housing sector.

Each house has two floors (ground and first floor) and the layout is similar: the kitchen and living area are on the ground floor with the bathroom and the bedrooms on the first floor. Further information on the houses under study can be found in Tables 1.1 and 1.2 as well as in Cuéllar-Franca and Azapagic (2012).

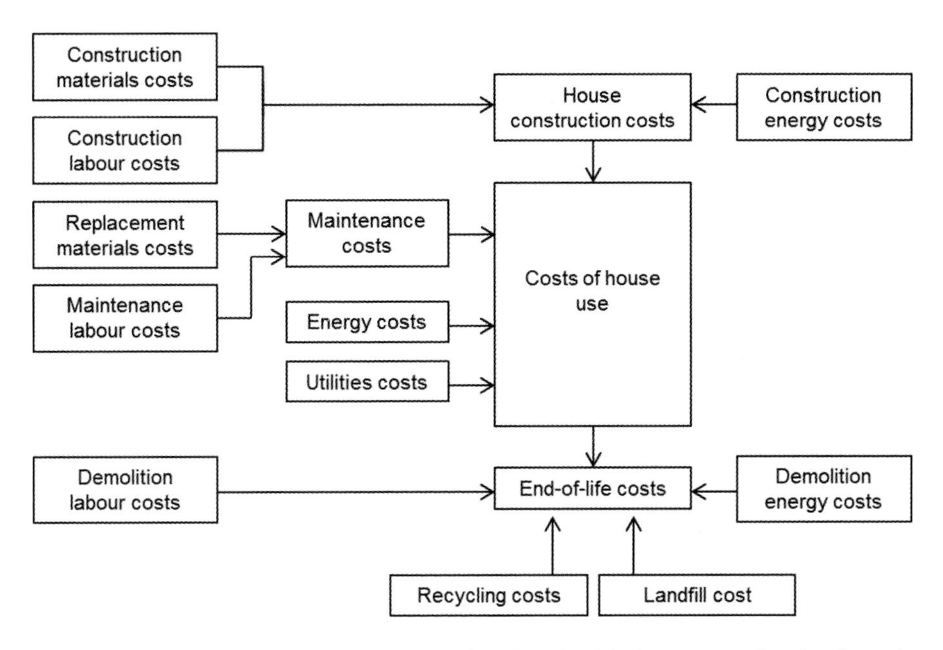

Figure 1.1 *System boundaries and the costs in the life cycle of the houses considered in the study.*

Table 1.1 *General information on the houses under study.*

	Detached house	*Semi-detached house*	*Terraced house*	*Source*
Usable floor area (m^2)	130	90	60	Brinkley (2008)
Household size (no. of people)	2.3	2.3	2.3	Utley and Shorrock (2008)
Number of bedrooms	4 bedrooms	3 bedrooms	2 bedrooms	Utley and Shorrock (2008)
Number of floors	2	2	2	BTP (2010)
Construction type	Traditional build: brick and block	Traditional build: brick and block	Traditional build: brick and block	BTP (2010)
Indoor temperature (°C)	19	19	19	Utley and Shorrock (2008); Brinkley (2008)
Air exchange rate (hr^{-1})	1	1	1	Brinkley (2008)
Specific heat loss (W/K)	220	170	120	Brinkley (2008); BTP (2010)

Table 1.2 *Materials used for the construction of houses.*

Element	*Surface (m^2)*			*Components*	*Thickness (mm)*	*Amount (kg)*		
	Detached	*Semi-detached*	*Terraced*			*Detached*	*Semi-detached*	*Terraced*
External wall	194	141	90	Brick (Imperial 9″), outer leaf	102.5	43,828	31,747	20,193
				Cement mortar	10	11,662	8,447	5,373
				Extruded polystyrene	75	510	292	96
				Concrete block (aerated), inner leaf	100	14,577	10,559	6,716
				Plasterboard	12.5	1,944	1,408	895
				Gypsum plaster skimming	3	653	473	301
Internal wall	99	85	44	Brick (Imperial 9″), inner leaf	102.5	22,302	19,199	9,809
				Cement mortar	10	5,934	5,108	2,610
				Plasterboard	12.5	1,978	1,703	870
				Gypsum plaster skimming	3	665	572	292
Foundation	30	25	19	Brick (Imperial 9″)	–	16,144	13,362	10,956
				Cement mortar	–	1,044	870	726
				Concrete	–	19,615	16,157	13,094
Ground floor	62	43	28	Cement mortar	20	190	233	173
Ceiling and first floor (bathroom and bedrooms)	4.4	5.4	4	Timber floor boards	20	640	443	288
	58	39	24	Carpet (bedrooms)	–	30	21	12
				Ceramic floor tiles (bathroom)	–	70	86	64
				Mineral wool	200	236	163	106
				Softwood timber (main beams and joists)	–	1,104	767	504

				Plasterboard	12.5	621	430	280
				Gypsum plaster skimming	3	209	144	94
Ground floor (kitchen, toilet and living room)	65	45	30	Ceramic floor tiles (kitchen/ toilet)	–	303	162	84
	25	10	5	Cement mortar	20	822	438	227
	40	35	25	Laminated floor (living room)	–	264	227	161
				Concrete slab	100	15,600	10,824	7,200
				Expanded polystyrene	100	150	104	69
				Damp-proof membrane	–	16	11	8
				Sand and gravel	50	7,280	5,051	3,360
Roof (timber structure)	75	52	35	Concrete tiles	–	3,750	2,602	1,732
	81	54	38	Sarking felt	–	9	7	4
				Softwood timber (purlins, ridge and wall plates, rafter, battens and truss membranes)	–	2,478	1,668	1,185
First floor ceiling	65	45	30	Softwood timber (joists)	–	78	54	38
				Mineral wool	300	449	311	207
				Plasterboard	12.5	650	451	300
				Gypsum plaster skimming	3	218	152	101
Windows	13	10	8	U-PVC frame	–	254	207	167
				Double glazed panes	–	197	160	129
Interior doors	11	9	6	Hardwood timber	34	292	250	167
Exterior doors	3	3	3	Hardwood timber	44	121	121	121
				Total materials (kg)		**176,931**	**134,965**	**88,701**

Data source: Courtesy of Cuéllar-Franca and Azapagic (2012).

1.2.2 Calculation of life cycle costs

The total life cycle costs of a house comprise the costs of construction, use and end-of life waste management and are calculated as follows:

$$LCC = C_{C} + C_{U} + C_{EoL} \tag{1.1}$$

where:

LCC – total life cycle costs of a house
C_{C} – costs of house construction
C_{U} – costs in the use stage of the house
C_{EoL} – costs of end-of-life of the house.

The construction costs C_{C} comprise the costs of the production and transport of construction materials as well as the labour and energy costs for the construction of the house and developer's profits:

$$C_{C} = C_{CM\&T} + C_{L\&OH} + C_{MF} + P_{D} \tag{1.2}$$

where:

$C_{CM\&T}$ – costs of extraction, production and transport of construction materials
$C_{L\&OH}$ – labour and overhead costs
C_{MF} – fuel costs for the machinery used in the construction of the house
P_{D} – developer's profits.

The costs incurred in the use stage comprise the costs of energy for space and water heating, lighting, cooking and domestic appliances as well as the costs for water and waste water treatment. Maintenance costs during its service life are also considered and include cost of labour, materials, energy and transport associated with the replacement of windows, doors and floor covering:

$$C_{U} = C_{E} + C_{W} + C_{M} \tag{1.3}$$

where:

C_{E} – costs of energy
C_{W} – costs of water and waste water
C_{M} – costs of maintenance.

Finally, end-of-life includes costs of labour and energy (fuel) for demolishing a house, waste collection services and transport costs as well as waste management costs such as landfilling and recycling of construction waste:

$$C_{EoL} = C_{DL} + C_{DF} + C_{WC} + C_{WT} + C_{LF} + C_{REC} \tag{1.4}$$

where:

C_{DL} – costs of labour for demolishing the house
C_{DF} – costs of fuel for the demolition machinery
C_{WC} – costs of collection of demolition waste
C_{WT} – costs of transport of demolition waste
C_{LF} – costs of waste landfilling
C_{REC} – costs of waste recycling.

In addition to the costs, there will be some revenue from selling the construction waste for reuse and/or recycling for which the system should be credited. However, due to a lack of data, this is excluded from the study but it is considered as a part of the sensitivity analysis later in this chapter.

All the costs represent 'overnight' costs, i.e. as if incurred at present time for all the life cycle stages so that no discounting is applied (BS, 2008, 2012). This is for two reasons. First, the economic performance of the houses is expressed in cost rather than financial-value terms. Second, one of the aims of this chapter is to identify cost hot spots in the life cycle of houses and related improvements, which would be applicable and carried out in the present time rather than in the future. However, the influence of more volatile costs such as energy on the total LCC is analysed as part of the sensitivity analysis as suggested by ISO 15686-5 (BS, 2008).

1.3 Life cycle costs of individual houses

The cost estimations for each life cycle stage are based on the data in Tables 1.2 and 1.3. The total costs are assessed from the supply chain perspective while the costs in each stage are estimated based on the perspective of the key player in that stage. The following section provides an overview of the costs by life cycle stage, followed by the overall life cycle costs in section 1.3.2 and costs per floor area in section 1.3.3. Note that the costs discussed below have been rounded off; the actual estimated values can be found in the corresponding figures and tables.

1.3.1 Costs by life cycle stage

1.3.1.1 Construction stage

The costs in this stage refer to the costs to the construction company or developer. The costs of construction materials and their transport as well as of energy and labour have been sourced from a construction cost guide for the UK (Hutchins, 2010). Land costs are excluded from the study as these are highly variable, depending on the location and land ownership models. This is also congruent with the recommendations in ISO 15686-5 (BS, 2008).

As detailed in Table 1.3 and summarised in Figure 1.2, the total construction costs are estimated at around £92,000 for the detached, £67,800 for the semi-detached and £44,000 for the terraced house, including a 10% profit for the construction company (Hutchins, 2010). The main contributors to the costs are the construction labour (52%) and materials (35%). The former is estimated at around £48,000 for the detached, £35,600 for the semi-detached and £23,000 for the terraced house, including a 20% overhead (Hutchins, 2010). The costs of the construction materials are £34,600 for the detached, £25,600 for the semi-detached and £16,700 for the terraced house. These costs include labour for the extraction and manufacture of the materials as well as the costs of transportation to the construction site. The breakdown for the materials costs can be seen in Figure 1.3.

As can also be seen from Figure 1.2, the total costs of fuels and machinery hire are small by comparison: £680 for the construction of the detached house, £470 for the semi-detached and £310 for the terraced house.

Table 1.3 *Costs of house construction (£).*

Activities/elements/ components	*Detached house*				*Semi-detached house*				*Terraced house*			
	Materials	*Labour*	*Fuel and machinery*	*Profit*	*Materials*	*Labour*	*Fuel and machinery*	*Profit*	*Materials*	*Labour*	*Fuel and machinery*	*Profit*
Construction activities			**564**				**390**				**260**	
External wall	**13,050**	**18,733**		**3,178**	**9,568**	**13,733**		**2,330**	**6,244**	**8,959**		**1,520**
Brick (Imperial 9″) and cement mortar	7,230	8,916		1,615	5,255	6,480		1,174	3,354	4,136		749
Extruded polystyrene	1,847	1,098		294	1,342	798		214	857	509		137
Concrete block	3,104	5,962		907	2,256	4,333		659	1,440	2,766		421
Plasterboard	704	993		170	595	840		143	516	729		125
Gypsum plaster skimming	165	1,763		193	120	1,282		140	77	818		89
Internal wall	**4,673**	**7,202**		**1,187**	**3,911**	**6,041**		**995**	**2,012**	**3,109**		**512**
Brick (Imperial 9″) and cement mortar	3,690	4,550		824	3,168	3,907		707	1,640	2,022		366
Plasterboard	899	1,269		217	671	947		162	335	472		81
Gypsum plaster skimming	84	1,383		147	72	1,187		126	37	615		65
Foundation	**1,577**	**1,636**		**321**	**1,396**	**1,433**		**283**	**1,137**	**1,169**		**231**
Double brickwork and cement mortar	732	899		163	606	744		135	497	610		111
Concrete	846	737		158	791	689		148	641	558		120
First floor	**3,260**	**5,130**		**839**	**2,423**	**3,880**		**630**	**1,508**	**2,440**		**395**

Carpet (polypropylene)	577	919	150	390	621	101	238	382	62
Ceramic floor tiles and cement mortar (Bathroom)	126	337	46	155	413	57	115	306	42
Timber floor boards (softwood)	1,009	1,036	205	698	717	142	455	467	92
Mineral wool	530	287	82	367	199	57	238	129	37
Joists (100 × 50 mm) – softwood timber	7	13	2	4	7	1	4	7	1
Main beams (200 × 50 mm) – softwood timber	767	1,161	193	641	970	161	349	529	88
Plasterboard	192	247	44	133	171	30	87	111	20
Gypsum plaster skimming	53	1,129	118	37	782	82	24	509	53
Ground floor	**2,433**	**5,050**	**748**	**1,605**	**3,384**	**499**	**1,027**	**2,197**	**322**
Ceramic floor tiles and cement mortar (kitchen/ toilet)	544	1,455	200	286	766	105	143	383	53
Laminated floor (oak)	147	1,892	204	112	1,440	155	80	1,029	111
Concrete slab	700	1,016	172	484	704	119	323	469	79
Expanded polystyrene	641	686	133	444	475	92	296	317	61
DP membrane (polypropylene)	30	0	3	21	0	2	14	0	1

(*Continued*)

Table 1.3 *Continued*

Activities/elements/ components	*Detached house*				*Semi-detached house*				*Terraced house*			
	Materials	*Labour*	*Fuel and machinery*	*Profit*	*Materials*	*Labour*	*Fuel and machinery*	*Profit*	*Materials*	*Labour*	*Fuel and machinery*	*Profit*
Sand	186	0		19	129	0		13	86	0		9
Gravel	186	0		19	129	0		13	86	0		9
Roof	**7,570**	**7,969**	**116**	**1,554**	**5,077**	**5,137**	**81**	**1,021**	**3,568**	**3,700**	**54**	**727**
Concrete tiles	3,819	2,695		651	2,648	1,868		452	1,782	1,258		304
Sarking felt	1,289	1,537	116	283	894	1,065	81	196	602	717	54	132
Purlins (75 × 225 mm)	170	129		30	94	71		16	68	51		12
Ridge plate (50 × 100 mm)	19	44		6	10	24		3	8	18		3
Wall plates (75 × 100 mm)	57	88		14	31	48		8	23	35		6
Rafter (50 × 100 mm)	326	755		108	227	526		75	162	375		54
Battens (25 × 38 mm)	281	1,062		134	141	531		67	129	488		62
Truss: Bottom membrane (50 × 200 mm)	424	655		108	295	456		75	210	325		54
Truss: Internal membrane (50 × 125 mm)	1,185	1,006		219	737	547		128	584	433		102

Ceiling	**957**	**1,472**		**243**	**663**	**1,019**		**168**	**443**	**682**		**112**
Joists (100 × 50 mm) – softwood timber	30	69		10	21	48		7	15	34		5
Mineral wool	671	364		104	465	252		72	310	168		48
Plasterboard	201	258		46	139	179		32	93	119		21
Gypsum plaster skimming	55	781		84	38	541		58	26	361		39
Windows	**452**	**742**		**119**	**359**	**592**		**95**	**245**	**402**		**65**
Bedroom window (1.1 m × 1.8 m)	268	444		71	201	333		53	134	222		36
Living room window (1.35 m × 1.6 m)	73	121		19	73	121		19	73	121		19
Kitchen window (1.1 m × 1.6 m)	59	99		16	59	99		16	12	20		3
Bathroom window (0.5 m × 0.7 m)	52	78		13	26	39		7	26	39		7
Interior doors	**298**	**300**		**60**	**255**	**257**		**51**	**170**	**171**		**34**
Exterior doors	**348**	**120**		**47**	**348**	**120**		**47**	**348**	**120**		**47**
Total	**34,618**	**48,354**	**680**	**8,364**[a]	**25,605**	**35,596**	**471**	**6,176**[a]	**16,702**	**22,949**	**314**	**3,996**[a]

Data sources: Courtesy of Hutchins (2010) and BBC (2011).
[a]Includes 10% profit for the total costs for the fuel and machinery.

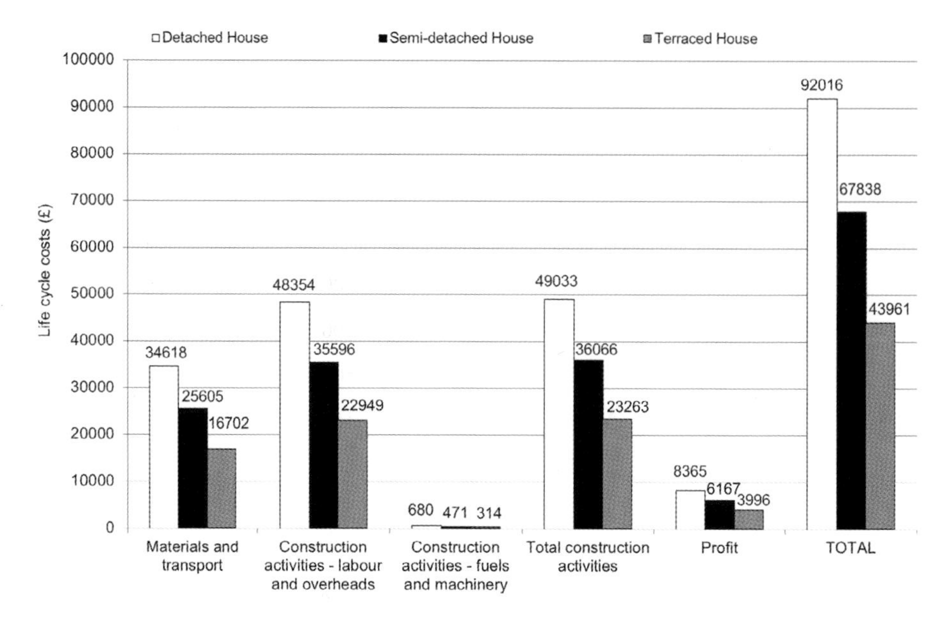

Figure 1.2 *Life cycle costs for the construction stage.*

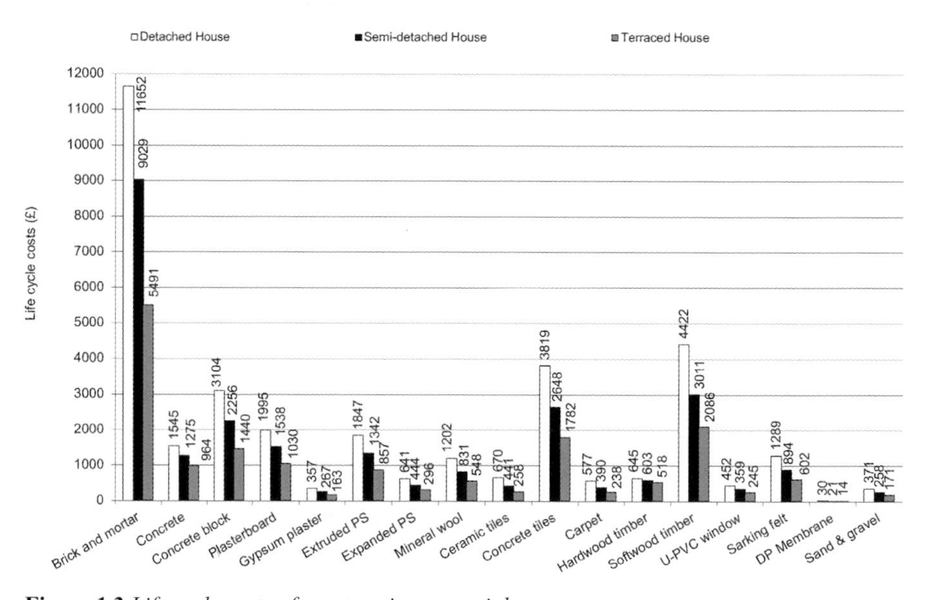

Figure 1.3 *Life cycle costs of construction materials.*

1.3.1.2 Use stage

The costs in the use stage are considered from the home owner or occupier perspective. These include the cost of energy for space and water heating, cooking, lighting and the use of domestic appliances (see Table 1.4). The energy consumption figures are based on the house floor area and household size defined in section 1.2.1.

Table 1.4 *Energy consumption for the three types of house over 50 years (MJ).*

Life cycle stage	*Detached house*	*Semi-detached house*	*Terraced house*
Construction			
On-site construction	*31,200*	*21,600*	*14,400*
Use			
Space heating	2,820,000	2 ,160,000	1,602,000
Water heating[a]	912,500	912,500	912,500
Cooking[a]	103,500	103,500	103,500
Lighting	255,750	151,150	93,700
Appliances[a]	314,360	314,360	314,360
Use total	*4,406,110*	*3,641,510*	*3,026,060*
End-of-life			
Demolition	*14,500*	*1, 000*	*6,700*
Total	**4,451,810**	**3,673,110**	**3,047,160**

Data source: Courtesy of Cuéllar-Franca and Azapagic (2012).
[a]Note that the amount of energy used for water heating, cooking and appliances is the same for all three types of house as these activities depend on the number of occupants which is equal for all three types.

The energy for space heating also takes into account the specific heat loss for each house, following the method suggested in Brinkley (2008). Water heating estimations are based on the water heating figures for the residential sector reported in Utley and Shorrock (2008), assuming a daily domestic hot water consumption of 76.5 litres per person (DCLG, 2010). The figures for cooking are derived from Utley and Shorrock (2008) and those for appliances and lighting are based on the figures reported in Baker and Jenkins (2007). Based on these figures, the total energy consumed over the lifetime of the houses ranges from 3.05 TJ for the terraced to 4.45 TJ for the detached house (Table 1.4).

The breakdown of the costs associated with water consumption and discharge can be found in Table 1.5. Assuming an average water consumption of 150 litres per day per person (DEFRA, 2010) and the breakdown of water usage by activity (MTP, 2008; VADO, 2010), the total water consumption is estimated at 6,280 m^3 over the lifetime of the house.

Therefore, the total running costs over the service life of a detached house are equal to £118,900 (Figure 1.4). For the semi-detached and terraced these costs are £98,000 and £81,300, respectively. The main contributors to the costs in this stage are energy consumption for space and water heating (40%) and maintenance (on average, 23%).

The total energy costs are between £47,200 and £67,200 (Table 1.6). This equates to the annual energy bill of £950 for the terraced house and £1,350 for the detached, giving an average energy bill of £1,135. This estimate is comparable to the average energy bill of £1,200 reported by Bolton (2010).

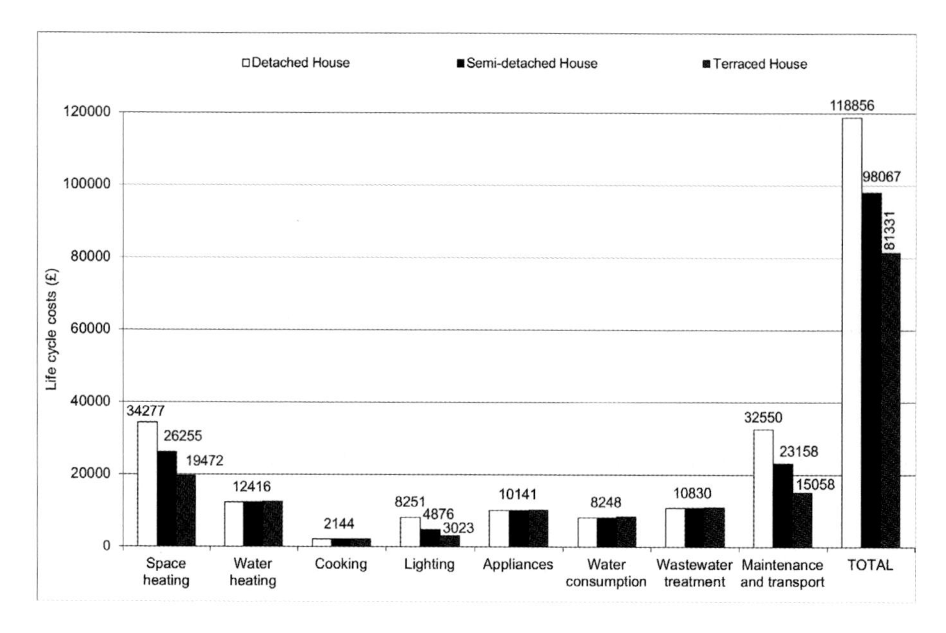

Figure 1.4 *Life cycle costs in the use stage.*

Table 1.5 *Breakdown of water consumption per house over 50 years*[a].

Activities	*End use (%)*	*Volume used (m³)*	*Volume discharged (m³)*
Personal hygiene	30	1,890	1,890
W.C.	30	1,890	1,890
Washing machine/dishwasher	21	1,320	1,320
Housekeeping	8	500	500
Personal consumption	4	250	0
Gardening	4	250	250
Others	3	200	200
Total	**100**	**6,300**	**6,050**

Data sources: Courtesy of Defra (2010), MTP (2008) and VADO (2010).
[a]Note that the total water consumption is equal for all three types of house as the number of occupants is the same.

The costs of water use and waste water treatment are estimated at £19,100 or around £380 annually (Table 1.6). These costs are the same for all three types of house owing to the same number of occupants, which determines the amount of water used (as opposed to energy which, for some types of use, e.g. heating, also depends on the size of the house).

Table 1.6 *Annual (and lifetime) energy and utility costs in the use stage.*

Fuel/energy source[a]	*Detached house*	*Semi-detached house*	*Terraced house*
Gas, £/yr (£)	684 (34,200)	564 (28,200)	462 (23,100)
Heating oil, £/yr (£)	87 (4,350)	71 (3,550)	57 (2,850)
Solid fuels, £/yr (£)	13 (650)	11 (550)	8 (400)
Electricity, £/yr (£)	560 (28,000)	471 (23,550)	416 (20,800)
Total annual energy costs, £/yr (£)	**1,344 (67,200)**	**1,117 (55,850)**	**944 (47,200)**
Water, £/yr (£)		165 (8,250)	
Waste water, £/yr (£)		217 (10,850)	
Total annual utility costs, £/yr (£)		**382 (19,100)**	

[a]Unit energy prices: gas: £10.84/GJ; oil: £13.67/GJ; solid fuels: £10.13/GJ; electricity: £32.28/GJ (Bolton, 2012).
Unit costs of water: £1.31/m^3; unit cost of waste water treatment: £1.72/ m^3 (OFWAT, 2010).

Table 1.7 *Total costs of maintenance and repairs over the lifetime of the house (£).*

	Detached house	*Semi-detached house*	*Terraced house*
Windows	1,693	1,355	1,751
Interior doors	1,459	1,250	1,592
Exterior doors	1,069	1,069	2,041
Floor coverings	28,330	19,483	23,365
Total	**32,550**	**23,158**	**15,059**

The maintenance costs include replacement costs for carpets (every five years), doors, ceramic and laminate flooring (every 20 years) and windows (25 years). Using the costs for the replacement materials and components given in Table 1.3, the overall costs of maintenance over the lifetime of the houses are estimated at around £32,550 for the detached house, £23,200 for the semi and £15,100 for the terraced house (see Table 1.7). These include the costs of labour, materials, energy as well as a 10% profit for the maintenance companies (Hutchins, 2010).

1.3.1.3 End-of-life stage

The costs in this stage are assessed from the perspective of the companies involved in different end-of-life activities such as demolition companies, waste collectors and landfill operators. The end-of-life materials and assumed waste management practices are given in Table 1.8; the assumed transportation distances for the waste materials are summarised in Table 1.9.

Table 1.8 *End-of-life materials and waste management options.*

	Reused/recycled/ landfilled (%)[a]	*Detached house (kg)*	*Semi-detached house (kg)*	*Terraced house (kg)*
Concrete, binders and aggregates	0/100/0	81,995	61,292	41,809
Bricks	51/36/13	82,281	64,295	40,950
Gypsum	0/100/0	6,939	5,331	3,133
Ceramic tiles	57/7/36	934	619	369
Insulation	18/0/82	1,360	882	485
Other inert waste (carpets, glass, etc.)	15/15/70	704	533	388
Timber	2/79/19	6.079	4,428	3,137
U-PVC	0/50/50	507	413	333
Total		**180,799**	**137,793**	**90,604**

Data sources: Courtesy of Kohler and Davies (2007), CRW (2009) and BBC (2011).
[a]Based on current UK waste management practice. Unit price of diesel for the demolition machinery: £1.33/l.

AQ1 **Table 1.9** *Transportation in the life cycle of the three types of house (t·km).*

	Detached house	*Semi-detached house*	*Terraced house*
Construction	8,400	6,350	4,000
Use (maintenance)	200	120	80
End of life	8,500	6,500	4,200
Total	**17,100**	**12,970**	**8,280**

Data source: Courtesy of Cuéllar-Franca and Azapagic (2012).

As shown in Table 1.10, the total costs of house demolition are around £16,600 for the terraced house, £26,400 for the semi-detached and £36,200 for the detached house. The majority of these (85%) are due to the labour costs, ranging from £13,700 to £30,500, respectively (Figure 1.5). The next highest cost item is waste collection from the demolition site (£2,150–£4,300). As indicated in Figure 1.5, the cost of fuel used for the demolition and waste disposal costs are small (in the order of several hundred pounds) while the costs of transport of the demolition waste are negligible (tens of pounds).

It should be noted that demolition companies usually offset the costs of demolition by selling construction waste as scrap (SilverCrest, 2010). Owing to a lack of data for the prices of recovered materials, it has not been possible to take this into account but a sensitivity analysis has been carried out (later in this chapter) to gauge the influence of the sales of second-hand materials on the total LCC costs.

Table 1.10 *End-of-life costs (£).*

	Detached house	*Semi-detached house*	*Terraced house*
Demolition energy (fuel)	500	346	231
Demolition labour	30,516	22,389	13,772
Demolition waste collection (incl. labour)[a]	4,306	2,990	2,133
Demolition waste transport (fuel)	64	48	32
Landfilling (incl. labour)[b]	528	400	256
Recycling (incl. labour)[c]	265	196	142
Total	**36,179**	**26,368**	**16,565**

Data sources: Courtesy of Hutchins (2010), SilverCrest (2010), Bolton (2010), OFWAT (2010), CRW (2009), Kohler and Davies (2007), BBC (2011) and Dewulf *et al.* (2009).
[a]Waste collection costs: £165 per 8 cubic yard skip of waste, including labour costs (SilverCrest, 2010).
[b]Landfill disposal cost: £25/t plus £2.50/t inactive waste and £56/t active waste (HMRC, 2011). Inactive waste includes bricks, aggregates, glass and mineral wool; active waste includes timber, UPVC and insulation materials.
[c]Fuel costs estimated using the fuel type and quantities specified for different machinery in EcoInvent (2011) and corresponding fuel prices (BBC, 2011; DECC, 2010). Labour costs for recycling estimated based on the time required to process the waste and the capacities of the recycling machinery (crusher, wood chipper and granulator) (MC, 2011; FF, 2011; PRC, 2011). Operator's wage: £20.85 per hour (Hutchins, 2010).

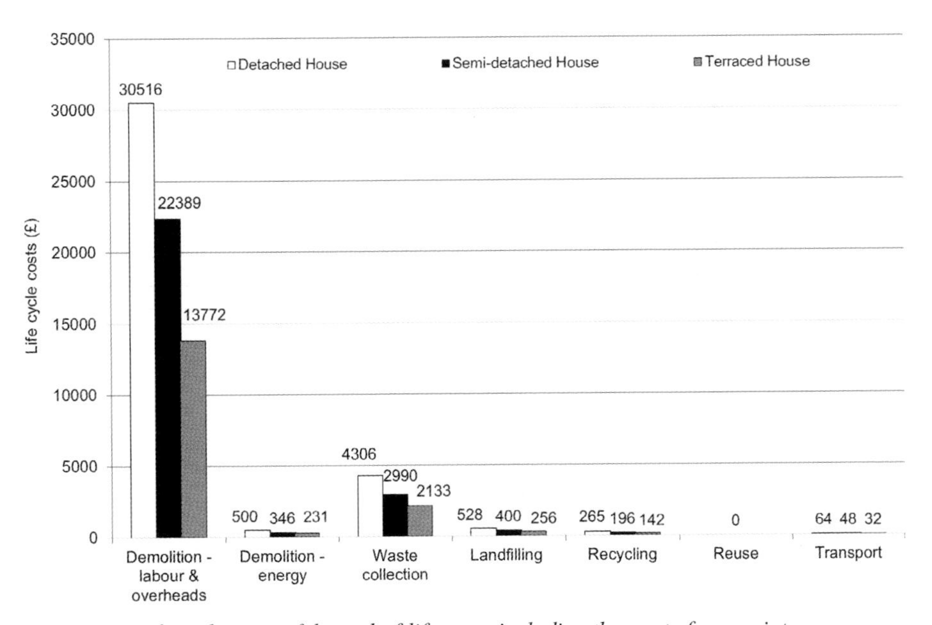

Figure 1.5 *Life cycle costs of the end-of-life stage including the waste from maintenance.*

1.3.2 Total life cycle costs and 'hot spots'

As shown in Figure 1.6, the total life cycle cost of a detached house is £247,000; for the semi-detached and terraced houses the equivalent values are £192,000 and £142,000, respectively. The use stage is responsible for the majority (52%) of the costs, largely from the use of energy. The construction stage contributes 35%, of which the labour costs contribute more than a half and construction materials around a third. Finally, end-of-life activities are responsible for 13% of the total costs, owing mainly to the labour costs.

Therefore, these areas should be targeted for reducing the housing life cycle costs. For example, making the existing houses more energy efficient through improved insulation, use of energy-efficient appliances and lighting would reduce energy bills but also environmental impacts since the use stage is also the hot spot for most impact categories (see Figure 1.7).

The second highest contributor to the LCC are the costs of construction materials such as bricks and concrete. Since these materials also have high embodied carbon (Cuéllar-Franca and Azapagic, 2012), both the costs and climate change impact of future housing stock could be reduced through material substitution but also more energy-efficient house designs. Some examples include timber-frame houses (Brinkley, 2008) as well as construction of 'passive' (BRE, 2011) and 'zero-carbon' houses (DCLG, 2010).

Finally, although the end-of-life costs are mainly due to the labour, demolition costs could be offset by increasing reuse and recycling of construction materials, which at the same time would reduce the amount of waste sent to landfill and related environmental

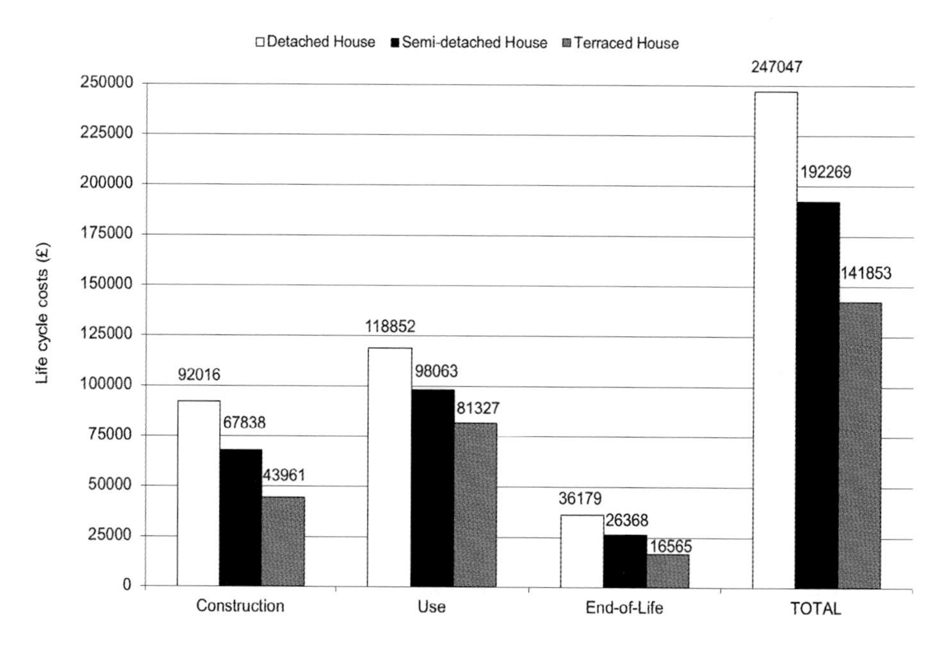

Figure 1.6 *Total life cycle costs for the detached, semi-detached and terraced house over the lifetime of 50 years.*

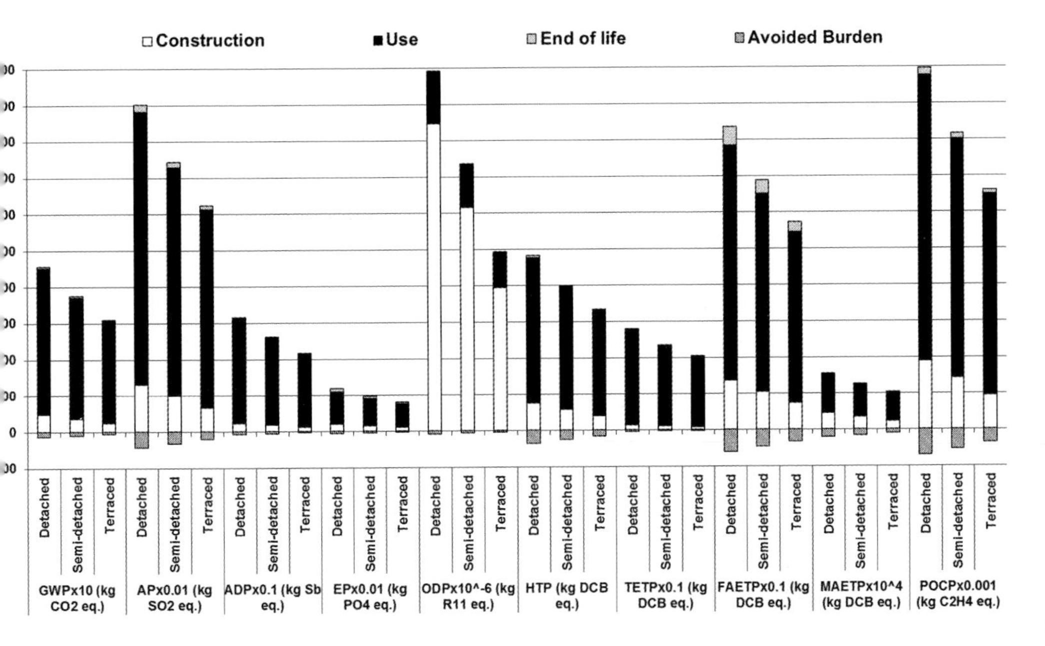

Figure 1.7 *Total life cycle impacts of the detached, semi-detached and terraced house over the lifetime of 50 years showing the contribution of different life cycle stages (Courtesy of Cuéllar-Franca and Azapagic, 2012).*

impacts. However, the potential for the latter from end-of-life activities is relatively small as their contribution to the total impacts is not significant (see Figure 1.7).

Further discussion on cost improvement opportunities at the sectoral level can be found in Section 1.4.

1.3.3 Life cycle costs per floor area

This section compares the costs of the three types of house per unit floor area to find out which house design may be more or less expensive overall. As can be observed from Figure 1.8, the detached house has the lowest life cycle costs (£1,900/m^2), followed by the semi-detached (£2,140/m^2) and the terraced house has the highest costs (£2,400/m^2). These differences are mainly due to the use stage as the construction and end-of-life costs are similar for the three designs. Given the same household size assumed for all three types of house and the fact that the energy used for water heating, cooking and appliances is the same (see Table 1.4), it is not surprising that the larger floor area (e.g. detached) will have a lower cost per unit area than a smaller one (e.g. terraced).

1.3.4 Validation of results

As already mentioned, there are no other studies of the LCC of houses in the UK so comparison of the results obtained here with previous studies is not possible. It is also not possible to compare the results obtained here with the studies based in Finland (Hasan *et al.*, 2008) and the USA (Keoleian *et al.*, 2000) because of the different methodologies used as well as different housing costs in these countries compared to the UK. For example, Hasan *et al.* (2008) do not consider full life cycle costs

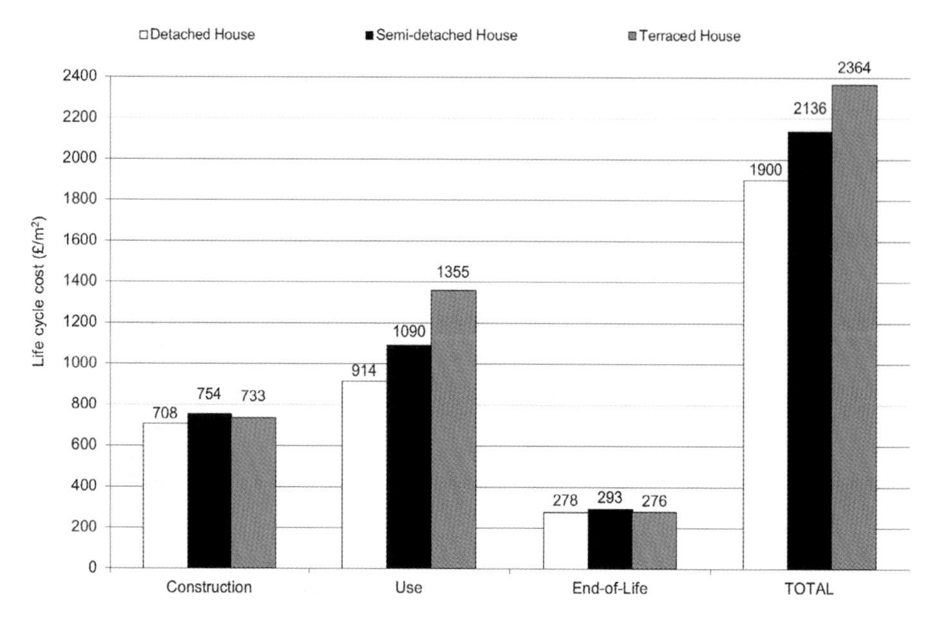

Figure 1.8 *Comparison of life cycle costs of the detached, semi-detached and terraced house per unit floor area over 50 years.*

because their focus is on the cost differential between a reference and several case studies. Keoleian *et al.* (2000), on the other hand, consider LCC from a homeowner perspective where construction costs are replaced by mortgage payments, rendering a comparison with the current study infeasible.

The only data that have been found for the UK are those related to the national statistics for the average household expenditure, which was estimated at around £2,600 in 2011 (Halifax, 2011). This is compared to the results obtained in this work in Figure 1.9. As shown, the results are relatively close (£1630–£2,380), despite the different assumptions and methodologies used.

1.3.5 Sensitivity analysis

1.3.5.1 Energy costs

This section considers the effect of potential future energy prices on the life cycle costs of houses. Three scenarios are considered and compared to the reference case. The latter assumes the 2010 energy prices fixed over the 50-year lifetime of the houses, as presented in section 3.1. Scenarios 1 and 2 assume a fixed annual increase of energy costs over the service life of a house, following the retail price index (RPI). Scenario 1 considers an annual fuel cost increase of 1.7% corresponding to the RPI between 2009 and 2010 and Scenario 2 assumes the increase of 2.5%, which corresponds to the cost increase between the first quarter of 2010 and first quarter of 2011 (DECC, 2011a). In Scenario 3, the energy cost increases annually by the average price increase between 2009 and 2011 (DECC, 2011a). During this period gas and electricity prices rose by 3.32% and 1.68%, respectively, while the cost of heating oils and coal increased by 3.83% and 6.12%, respectively.

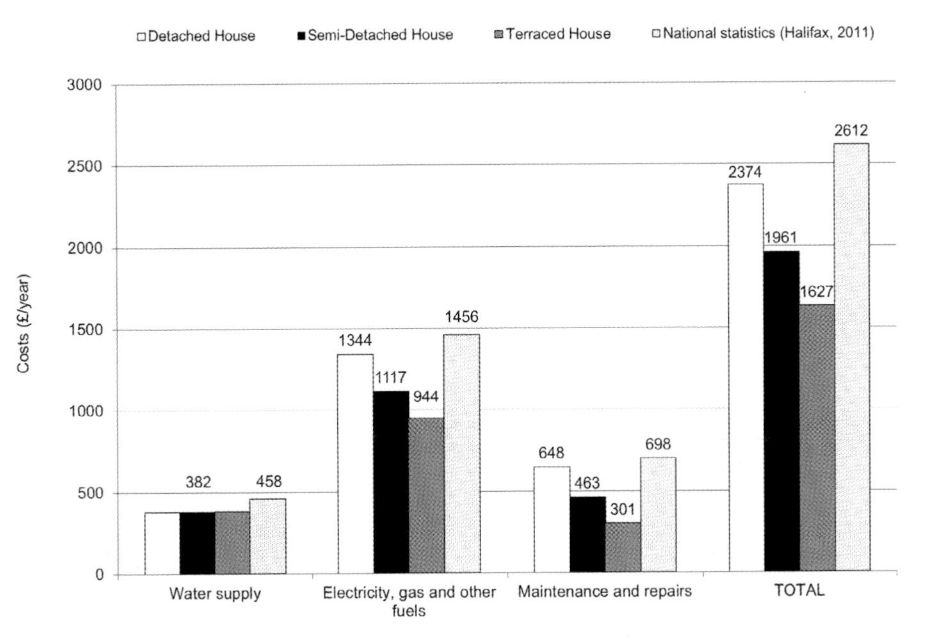

Figure 1.9 *Validation of the results: comparison of the costs from the use stage with the annual household expenditure.*

The total energy costs for the three types of house are shown in Figures 1.10–1.12. As indicated, the costs increase over the lifetime of houses by 55% in Scenario 1, 94% in Scenario 2 and 114% in Scenario 3 compared to the reference case.

Figure 1.13 shows the annual energy cost increments over the service life of the house for each scenario (for illustration, only the results for the detached house are shown). As indicated, Scenario 1 is the best and Scenario 3 the worst. According to the latter, by 2035 the energy bills could be two times higher than today and by 2050 they could triple. Thus the impact on the total life cycle costs of houses would be significant, making it more difficult to eradicate fuel poverty of UK households.

1.3.5.2 Reuse of end-of-life materials

The influence on the total LCC of revenue from the reuse of the end-of-life materials is illustrated in Figure 1.14. In the absence of real data, two scenarios are considered: one whereby the salvaged materials and components are sold back at the original price of the virgin materials (see Table 1.3) and another where only 50% of these costs are recovered. As indicated in the figure for the detached house, for example, around £9,400 can be recovered in the first scenario and a half of that in the second, reducing the total LCC by 3.8% and 1.9%, respectively. Thus, the influence of this parameter on the total LCC is insignificant.

1.3.5.3 Number of occupants

In this work, it has been assumed that all three houses have the same number of occupants (2.3 people), equal to the average size of the household in the UK. To find out

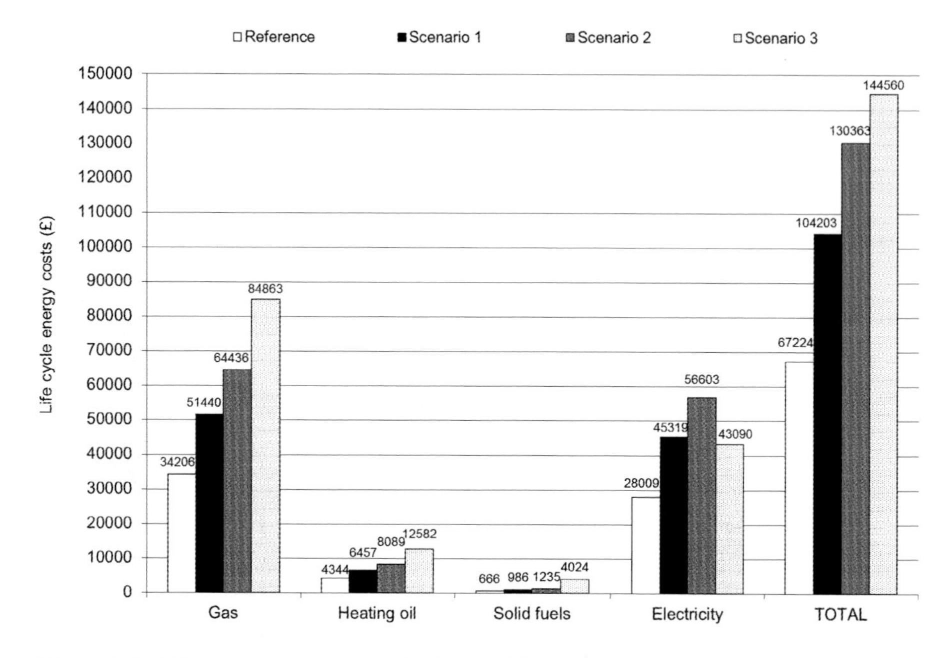

Figure 1.10 *Lifetime energy costs for the detached house for different scenarios.*

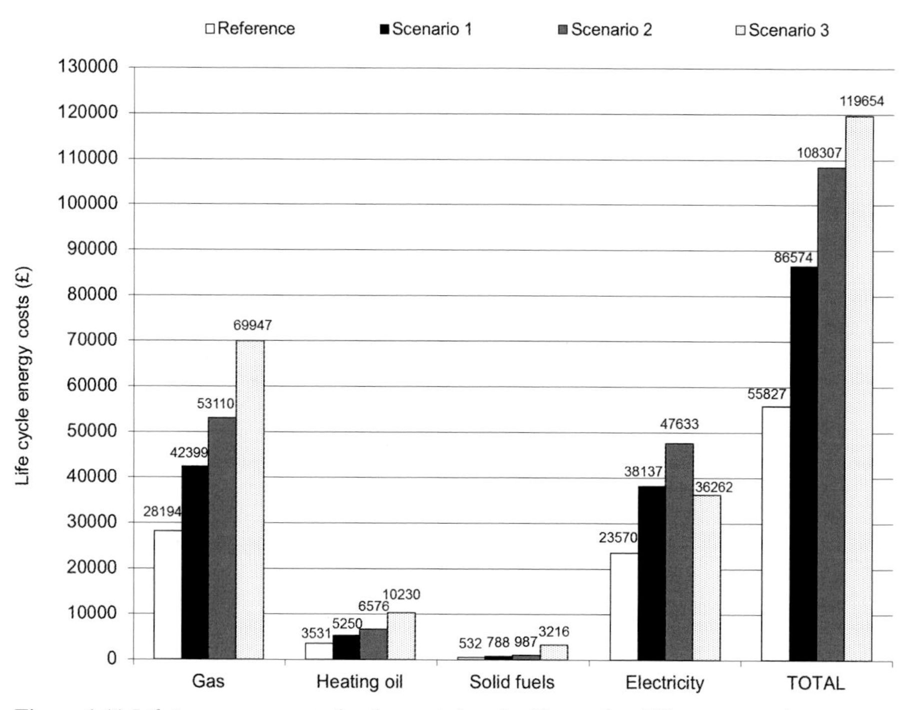

Figure 1.11 *Lifetime energy costs for the semi-detached house for different scenarios.*

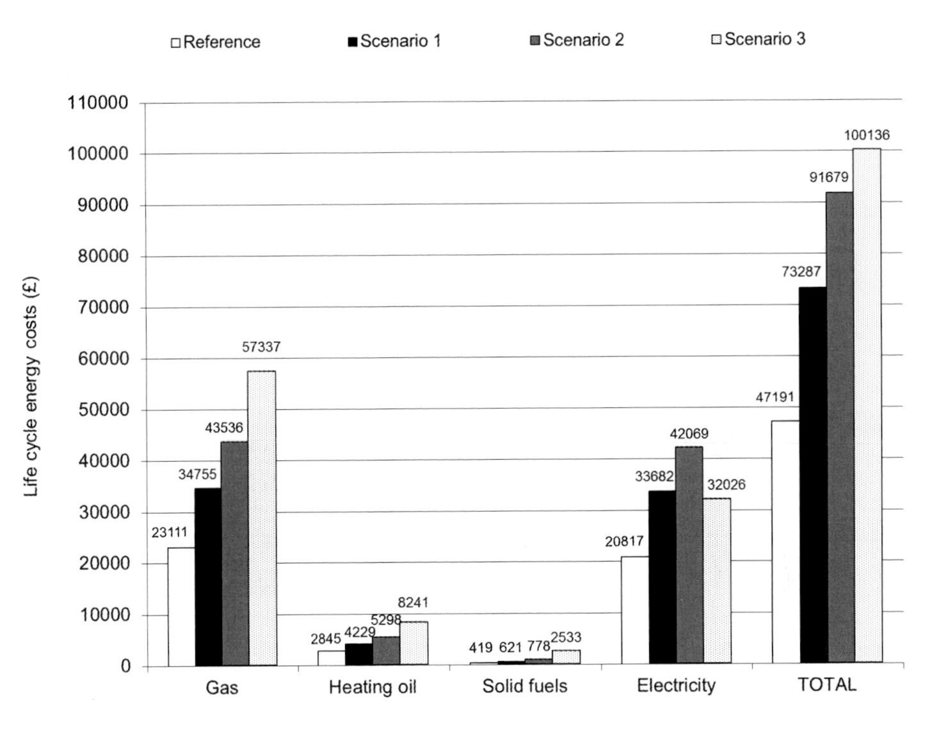

Figure 1.12 *Lifetime energy costs for the terraced house for different scenarios.*

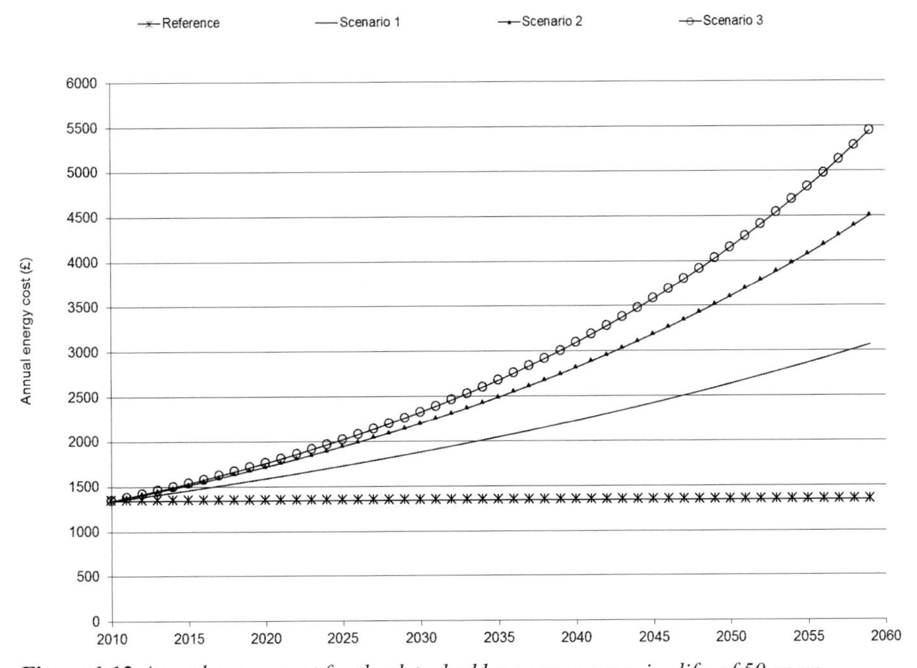

Figure 1.13 *Annual energy cost for the detached house over a service life of 50 years.*

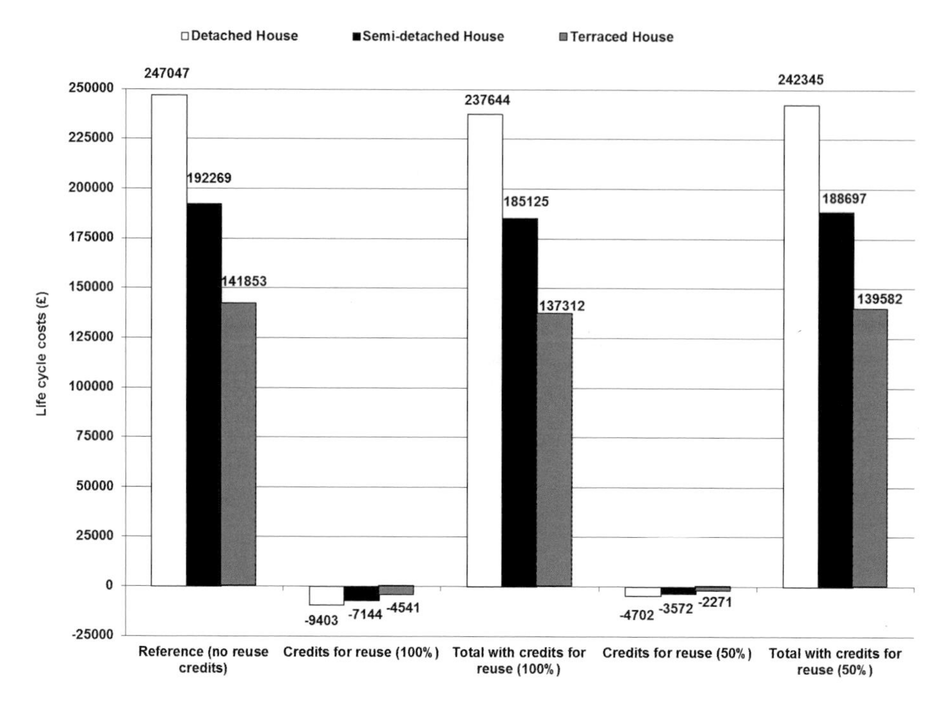

Figure 1.14 *Influence on the total life cycle costs of reusing the end-of-life materials (credits for reuse: 100% and 50% represent, respectively, the percentage of costs recovered compared to the cost of the virgin materials. For the materials assumed to be reused see Table 1.8; the costs of virgin materials are given in Table 1.3).*

how the number of occupants would influence the housing life cycle costs, a sensitivity analysis has been performed considering a theoretical number of occupants in each house based on their respective number of bedrooms, i.e. four people in the detached, three in the semi-detached and two in the terraced house.

These results are shown in Figure 1.15 and indicate that the number of occupants has a relatively small influence on the life cycle costs of a house. This is mainly because space heating accounts for the large majority of the operational costs and this depends on the usable floor area of the house rather than the number of occupants. For example, the life cycle costs of a detached house occupied by four people increase by 10% compared to the average household size of 2.3 people. For a semi-detached house with three people, the total life cycle costs increase by 5% and for the terraced house, the costs are reduced by 3% as the number of occupants is now lower (two) than originally assumed (2.3).

1.4 Life cycle costs of the existing housing stock in the UK

The LCC for the individual houses presented in the previous section have been extrapolated to the existing UK housing stock of seven million each of semi-detached and terraced and four million of detached houses (Utley and Shorrock, 2008). The results in Figure 1.16 indicate that the costs associated with the semi-detached houses are

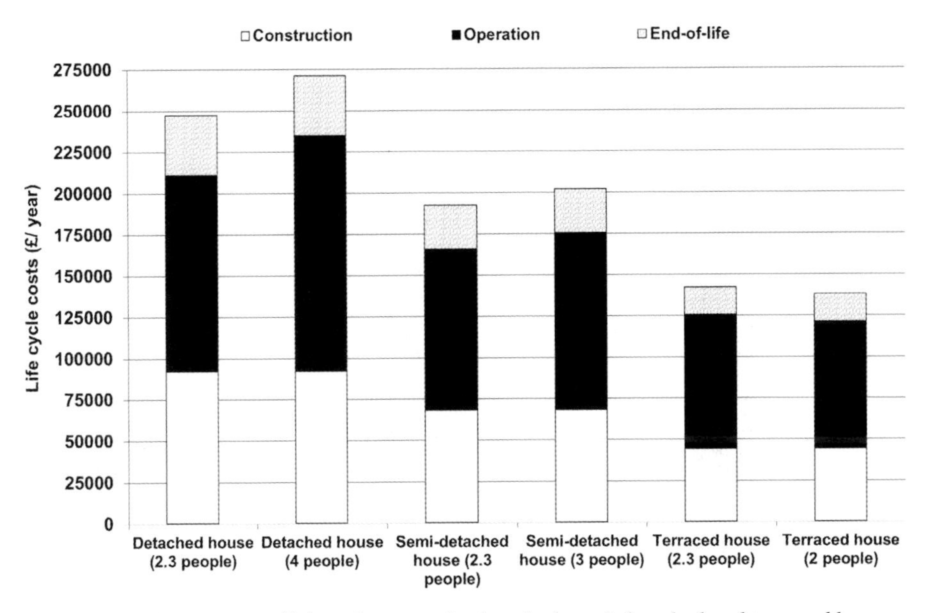

Figure 1.15 *Comparison of life cycle costs of a detached, semi-detached and terraced house for different number of occupants.*

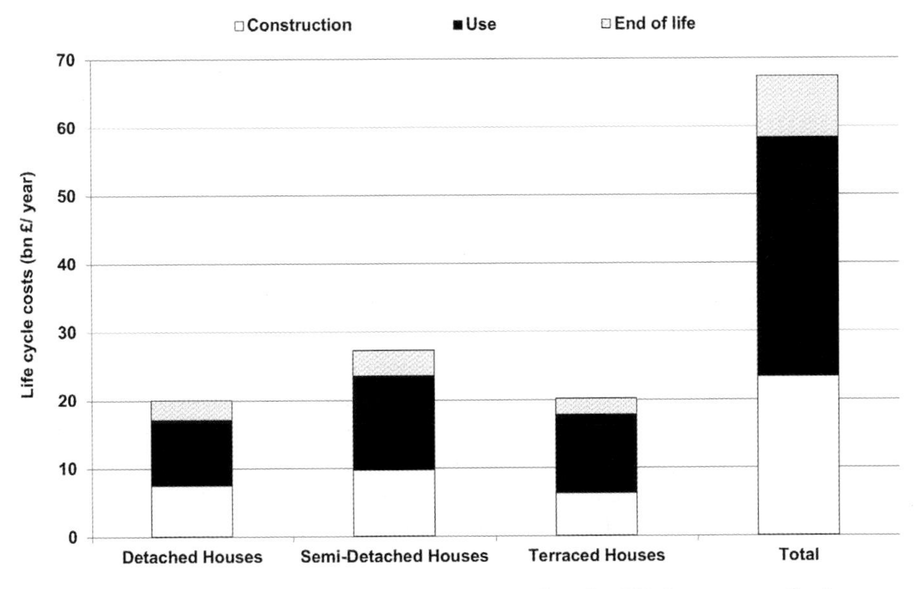

Figure 1.16 *Life cycle costs of the existing housing stock in the UK showing contributions from the life cycle stages.*

£27 billion per year while those of detached and terraced houses are £20 billion each. This gives a total of £67 billion per year or £3,364 billion over the 50-year lifetime. Although it is difficult to put these figures in context as there are no other comparable results, as an illustration, the UK's GDP in 2011 was £1.5 trillion (World Bank, 2012). Thus the current housing sector is 'worth' around 45 times the GDP.

Figure 1.16 also shows that around a half of the housing costs are generated during the use stage, of which around 50% or £990 billion (over 50 years, at 2010 prices) is from energy use.

The high dependency on fossil fuels makes the housing sector vulnerable to future increase of energy prices. As shown in section *Energy costs*, the energy costs are likely to increase well beyond the optimistic assumption of constant energy prices made here and could triple to reach around £3,000 billion by 2050. This is likely to increase the fuel poverty of UK households further (DECC, 2011b). Thus, in addition to improving energy efficiency, the housing sector will need to seek ways to become less dependent on fossil fuels and the volatility of energy prices. Some examples include installation of renewable technologies such as solar thermal panels, heat pumps and photovoltaics. However, this will require significant capital investments, pushing up the overall housing costs. These would need to be assessed carefully and balanced against the threat of the rising energy costs. The UK government has already introduced a range of incentives to stimulate installation of renewable technologies in the existing housing sector, including the feed-in-tariffs (FITs) (DECC, 2012a) and renewable heat incentives (RHI) (DECC, 2012b). However, due to a number of reasons, the uptake is still low providing less than 0.2% of the final energy demand in the UK domestic sector (Balcombe *et al.*, 2013, DECC, 2012c). This is largely due to the initial investment required, which is beyond the means of most UK households (Balcombe *et al.*, 2013).

In addition to the FITs and RHI, the Government has spent more than £30 billion on various measures to reduce fuel poverty over the last 11 years (2000–2011), including winter fuel payments, decent homes and warm front while energy companies were expected to spend £3.9 billion over past 3 years (2008–2011) on energy efficiency and social assistance for households (Bolton, 2010). This means that the Government and industry have collectively spent on various fuel poverty measures 15% of £220 billion spent by the householders on energy bills over the same period[1] – and yet six million households still live in fuel poverty, the same as in the 1990s (DECC, 2009; Bolton, 2010).

Thus, the Government will need to seek further opportunities and options for reducing fuel poverty. This includes improving the uptake of renewable technologies but also incentivising householders to improve energy efficiency of homes through improved insulation, energy-efficient appliances and lighting, etc. For example, some estimates indicate that it would cost around £3,000 per house to improve the energy efficiency, typically through loft and cavity wall insulation or a condensing boiler (EST, 2010). However, further measures and incentives will be needed to help reduce energy demand – while renewable technologies and fuel payments can certainly help towards reducing the dependency on fossil fuels and the costs of energy bills, it is the householders' behaviour that could have a much more significant impact on reducing the energy consumption and related costs.

The next largest contributor to the life cycle costs is house construction, representing 35% or £1,177 billion of the total housing costs. This is another issue affecting

[1]Calculated based on the annual energy costs given in Table 1.4 and extrapolated over 11 years for the total number of detached, semi-detached and terraced houses.

the residential construction sector as there is a stark difference between the construction costs and house market prices. The average house price reported in 2011 was £160,000 (HPUK, 2012) while the average construction cost estimated in this study is £68,000. Thus, the average market price is around 2.4 times higher than the average construction costs, despite that fact that the latter already include developers' profits. This gap will deepen as the economy starts to recover and the house prices go up which, together with expensive mortgage products, will inevitably lead to a further decline in home ownership, particularly affecting the prospects of young people to purchase a home. Therefore, unless appropriate measures are taken by the Government and other players in the construction sector, there is a danger of turning a basic human need – shelter – into a luxury commodity.

1.5 Conclusions

The total house life cycle costs over the 50-year lifetime range from £247,000 for the detached to £142,000 for the terraced house. Half of the costs are from the use stage of which around 50% is due to the costs of energy. This is mainly because of the use of gas for space and water heating and electricity costs. The average annual energy costs per household are estimated at £1,135. If the energy costs continue to grow as expected, in 2035 the energy bills could be twice as high and in 2050 they could triple. Thus the impact on the total life cycle costs of houses would be significant, making it more difficult to eradicate fuel poverty of UK households.

The construction costs contribute 35% to the total LCC of with the walls and the roof being the most expensive items. The remaining 13% of the costs are incurred at the end of life of the house which are largely (85%) due to the cost of labour for demolition. Recovery of end-of-life materials has a limited potential to reduce the overall life cycle costs of a house.

At the sectoral level, the total life cycle costs of housing are estimated at £67 billion per year with semi-detached houses contributing £27 billion, followed by the detached and terraced houses with £20 billion each. Over the 50-year lifetime, the total life cycle costs add up to £3,364 billion.

The existing housing stock in the UK is facing a number of challenges that will need to be addressed in the near future. These include improving energy efficiency through better insulation and use of energy-efficient appliances and lighting, which would reduce energy bills but also environmental impacts. Furthermore, reducing the dependency on fossil fuels could help to reduce both fuel poverty and environmental impacts. Moreover, the disparity between the construction costs and house market prices will need to be addressed to ensure that access to and house ownership do not become the privilege of a few.

Acknowledgements

This chapter was originally published by Springer as the following research paper: Cuéllar-Franca, R. M. and A. Azapagic (2013) Life cycle cost analysis of the UK housing stock. *The International Journal of Life Cycle Assessment,* **19**, 174–193. Reproduced here with kind permission of Springer Science+Business Media. One of the authors (RCF) would also like to thank CONACYT for their financial support.

References

Abeysundra, U.G.Y., B. Sandhya, G. Shabbir and S. Alice (2007) Environmental, economic and social analysis of materials for doors and windows in Sri Lanka. *Building and Environment* **42**(5), 2141–2149.

Adalberth, K. (1997) Energy use during the life cycle of single-unit dwellings: examples. *Building and Environment* **32,** 321–329.

Adalberth, K., A. Almgren and E.H. Petersen (2001) Life cycle assessment of four multi-family buildings. *International Journal of Low Energy and Sustainable Buildings* **2,** 1–21.

Asif, M., T. Muneer and R. Kelley (2007) Life cycle assessment: a case study of a dwelling home in Scotland. *Building and Environment* **42,** 1391–1394.

Balcombe, P., D. Rigby and A. Azapagic (2013) Motivations and barriers associated with adopting microgeneration energy technologies in the UK. *Renewable and Sustainable Energy Reviews* **22**(2013), 655–666.

Barker, T. and K. Jenkins (2007) The domestic energy sub-model in MDM-E3. http://www.ukerc.ac.uk/Downloads/PDF/07/0705ESMDomEnSub-Model.pdf. Accessed 02/06/14.

BBC (2011) *Diesel price reaches record high.* www.bbc.co.uk/news/business-12309329. Accessed 03/6/14.

BBC (2012) *Home ownership in england continues gradual decline BBC News.* www.bbc.co.uk/news/business-17026462. Accessed 04/06/14.

Bolton, P. (2010) Energy price rises and fuel poverty. *Key Issues for the New Parliament 2010,* House of Commons Library Research. www.parliament.uk/documents/commons/lib/research/key_issues/Key%20Issues%20Energy%20price%20rises%20and%20fuel%20poverty.pdf. Accessed 03/06/14.

Bolton, P. (2012) *Energy prices.* Last updated: 2 Nov 2012. House of Commons. www.parliament.uk/briefing-papers/SN04712.pdf. Accessed 03/06/14.

Bribián, I.Z., A.A. Usón and S. Scarpellini (2009) Life cycle assessment in buildings: state of the art and simplified LCA methodology as a complement for building certification. *Building and Environment* **44,** 2510–2520.

Brinkley, M. (2008) *The housebuilder's Bible*, 7th edition. Ovolo Publishing, UK.

BRE (2011) *Passivhaus primer: introduction (an aid to understand the key principles of the passivhaus standard).* Building Research Establishment, London. http://www.passivhaus.org.uk/filelibrary/Primers/KN4430_Passivhaus_Primer_WEB.pdf. Accessed 03/06/14.

BS (2008) BS ISO 15686-5:2008. Buildings & construction assets – service life planning – Part 5: Life cycle costing. British Standards.

BS (2012) BS EN 15643-4:2012. Sustainability of construction works – assessment of buildings. Part 4: Framework for the assessment of economic performance. British Standards.

BTP (2010) *Blue print for greenheys great western street moss side. Elevations – setting out block 8.* Bernard Taylor Partnership Ltd., Manchester.

CRW (2009) *Management of non-aggregated waste.* Construction, Resource and Waste Platform, UK.

Cuéllar-Franca, R.M. and A. Azapagic (2012) Environmental impacts of the UK residential sector: life cycle assessment of houses. *Building and Environment* **54,** 86–99.

DCLG (2008) *English housing survey: housing stock report 2008.* Department of Communities and Local Government. https://www.gov.uk/government/uploads/system/uploads/attachment_data/file/6703/1750754.pdf. Accessed 03/06/14.

DCLG (2010) *Code for sustainable homes: technical guide.* Department of Communities and Local Governments. Available at: http://www.planningportal.gov.uk/uploads/code_for_sustainable_homes_techguide.pdf Accessed 03/06/14.

DCLG (2012) *English housing survey: households report 2010–11*. Department of Communities and Local Government. www.communities.gov.uk/publications/corporate/statistics/ehs201011householdreport. Accessed 03/06/14.

DECC (2009) *The fuel poverty strategy*. Department of Energy and Climate Change, London.

DECC (2010) *Quarterly energy prices*. Department of Energy and Climate Change, London.

DECC (2011a) Retail price index: fuel components, monthly figures. www.decc.gov.uk/assets/decc/statistics/source/prices/qep213.xls. Accessed 03/06/14.

DECC (2011b) *Annual report on fuel poverty statistics 2011*. Department of Climate Change, London.

DECC (2012a) *Feed-in-tariffs*. Department of Energy and Climate Change, London. www.decc.gov.uk/en/content/cms/meeting_energy/renewable_ener/feedin_tariff/feedin_tariff.aspx. Accessed 03/06/14.

DECC (2012b) *Renewable heat incentive (RHI) scheme*. Department of Energy and Climate Change, London. www.decc.gov.uk/en/content/cms/meeting_energy/renewable_ener/incentive/incentive.aspx. Accessed 03/06/14.

DECC (2012c) *Feed-in-tariffs scheme. Government response to consultation on comprehensive review phase 2A: solar PV cost control*. Department of Energy and Climate Change, London.

DEFRA (2010) *Sustainable consumption and production: domestic water consumption*. Department for Environment, Food and Rural Affairs. Available at: http://archive.defra.gov.uk/sustainable/government/publications/uk-strategy/documents/Chap£.pdf. Not now (03/06/14) available.

Dewulf, J., N. Van der Vorst, Versele, A. Janssens and H. Van Langenhove (2009) Quantification of the impact of the end-of-life scenario on the overall resource consumption for a dwelling house. *Resources, Conservation and Recycling* **53**(4), 231–236.

EcoInvent (2011) Ecoinvent Centre, Zurich. www.ecoinvent.ch. Accessed 03/06/14.

EST (2010) *F & G banded homes in Great Britain – research into costs of treatment*. Energy Saving Trust, London. www.energysavingtrust.org.uk/Publications2/Housing-professionals/Refurbishment/F-G-banded-homes-in-Great-Britain-research-into-costs-of-treatment. Accessed 03/06/14.

Eurostat (2012) *European Union statistics on income and living standards*. European Commission.http://epp.eurostat.ec.europa.eu/portal/page/portal/income_social_inclusion_living_conditions/data/main_tables. Accessed 03/06/14.

FF (2011) *Wood chipper hire*. www.forestfuels.co.uk/wood-chipper-hire.php. Not now (03/06/14) available.

Gluch, P. and H. Baumann (2004) The life cycle costing (LCC) approach: a conceptual discussion of its usefulness for environmental decision-making. *Building and Environment* **39**, 571–580.

Hacker, J.N., T.P. De Saulles, A.J. Minson and M.J. Holmes (2008) Embodied and operational carbon dioxide emissions from housing: a case study on the effects of thermal mass and climate change. *Energy and Buildings* **40**(3), 375–384.

Halifax (2011) *Cost of housing rises to three year high*. www.lloydsbankinggroup.com/media/pdfs/halifax/2011/Cost_of_housing_rises_to_three_year_high.pdf. Accessed 03/06/14.

Hammond, G. and C. Jones (2008) Embodied energy and carbon of construction materials. *Proceedings of the ICE – Energy* **161,** 87–98.

Hasan, A., M. Vuolle and K. Sirén (2008) Minimisation of life cycle cost of a detached house using combined simulation and optimisation. *Building and Environment* **43**(12), 2022–2034.

HC (2008) *Construction matters. House of commons business and enterprise committee*. Http://www.publications.parliament.uk/pa/cm200708/cmselect/cmberr/127/127i.pdf. Accessed 03/06/14

HMRC (2011) *A general guide to landfill tax*. HM Revenue & Customs, London.

HPUK (2012) *House prices*. www.houseprices.uk.net/graphs. Not now (03/06/14) available.

Hunkeler, D., K. Lichtenvort and G. Rebitzer (2007) *Environmental life cycle costing*. CRC Press.
Hutchins (2010) *UK building blackbook: the small works and maintenance construction cost guide*. Vol. 1., Franklin and Andrews Ltd., London.
ISO (2006a) ISO 14040 – *Environmental management – life cycle assessment – principles and framework*. Geneva.
ISO (2006b) ISO 14044 – *Environmental management – life cycle assessment – requirements and guidelines*. Geneva.
Keoleian, G.A., S. Blanchard, and P. Reppe (2000) Life-cycle energy, costs, and strategies for improving a single-family house. *Journal of Industrial Ecology* **4**(2), 135–156.
Kohler, N. and S. Davies (2007) *Demolition exemplar case study: recycling demolition arisings at Hamilton House, Sandwell*. Waste Resource Action Programme, Oxon. www.wrap.org.uk/construction/case_studies/recycling_demolition.html. Not now (03/06/14) available.
MC (2011) *PE series jaw crusher*. www.zenithcrusher.com/products/crushing/jaw-crusher/capacity.html. Not now (03/06/14) available.
Monahan, J. and J.C. Powell (2011) An embodied carbon and energy analysis of modern methods of construction in housing: a case study using a lifecycle assessment framework. *Building and Environment* **43,** 179–188.
MTP (2008) Domestic water consumption in domestic and non-domestic properties. Market Transformation Programme. http://efficient-products.defra.gov.uk/spm/download/document/id/669. Not now (03/06/14) available.
OFWAT (2010) *Annual report 2010–11*. Water Services Regulation Authority. http://www.ofwat.gov.uk/aboutofwat/reports/annualreports/rpt_ar2010-11print.pdf. Accessed 03/06/14.
Ortiz, O., C. Bonnet, J.C. Bruno, and F. Castells (2009) Sustainability based on LCM of residential dwellings: a case study in Catalonia, Spain. *Building and Environment* **44**, 584–594.
Peuportier, B.L.P. (2001) Life cycle assessment applied to the comparative evaluation of single family houses in the French context. *Energy and Buildings* **33,** 443–450.
POST (2003) *Modern methods of house building*. Parliamentary Office of Science and Technology. Avaialble at: http://www.parliament.uk/briefing-papers/POST-PN-209.pdf. Accessed 03/06/14.
PRC (2011) *Plastic recycling*. www.plasticrecyclingcentre.co.uk/plasticrecyclingcentre. Accessed 03/06/14.
Rebitzer, G., D. Hunkeler, and O. Jolliet (2003) LCC – the economic pillar of sustainability: methodology and application to wastewater treatment. *Environmental Programme* **22**(4), 241–249.
RICS (2012) *RICS UK housing market survey*. Royal Institution of Chartered Surveyors. www.rics.org/site/scripts/documents_info.aspx?categoryID=409&documentID=34&pageNumber=2. Accessed 03/06/14.
Sergeant, M. (2011) *Home ownership set to fall in UK*. BBC News. www.bbc.co.uk/news/business-14711308. Accessed 03/06/14.
SilverCrest (2010) *Disposal of construction waste*, R.C. Franco, ed. Derbyshire.
Swarr, T.E., Hunkeler, D., Klöpffer, W., Pesonen, H-L., Ciroth, A., Brent, A.C. and Pagan, R. (2011) Environmental life-cycle costing: a code of practice. *International Journal of Life-cycle Assessment* **16**, 389–391.
UKCG (2009) *Construction in the UK economy: the benefits of investment*. UK Contractors Group, London.
Utley, J. and L. Shorrock (2008) *Domestic energy fact file*. Building Research Establishment, London.
VADO (2010) *Grey water recycling and rain water harvesting*. VADO and Hydrocyc UK. http://www.vado-uk.com/downloads/hydrocyc_2010.pdf. Not now (03/06/14) available.
World Bank (2012) *UK GDP for 2011*. http://data.worldbank.org/country/united-kingdom. Accessed 02/06/2014.

Chapter 2
Case study: Life cycle analysis of a community hydroelectric power system in rural Thailand

Andrew Pascale, Tania Urmee and Andrew Moore

2.1 Introduction

A 2003 report published by the United Nations Development Programme found a fundamental link between the Millennium Development Goals, which include halving extreme poverty, halting the spread of HIV/AIDS, enrolling all boys and girls everywhere in primary school by 2015 (UNDP, 2003, v) and the provision of energy services – especially in rural and developing country locations. However, only recently has a global initiative led by the United Nations (SE4ALL) set a 2030 target of universal access to modern energy services reaching among others, the 1.2 billion people that still lack access to electricity (World Bank/ESMAP and International Energy Agency 2013). Electricity is an important component of modern energy services and electricity access provides an indication of a population's standard of living (Kammen and Kirbui 2008).

Provision of electricity must be appropriate for the people, place, planet and time - especially in remote rural areas which are difficult to access and where local expertise is limited. Electricity must be affordable and sustainable – socially, economically, technically, and environmentally (Urmee, Harries, and Schlaprfer 2009). A range of off-grid options, like solar PV home system, small hydro, and other renewable options make it possible to provide the basic electricity needs of households, local communities and small businesses in rural areas where grid electricity is not an option in the foreseeable future (Urmee and Harries 2009). Doubling renewable energy's 2010 share of total final energy consumption by 2030 is another goal of the SE4ALL initiative (World Bank/ESMAP and International Energy Agency 2013).

In order to determine the appropriateness of a small energy solution for any location, the full environmental impacts of proposed energy solutions should be understood by all stakeholders. Solar photovoltaic (Alsema 2000, Moore, 2009), wind (Fleck and Hout, 2009) and biodiesel (Gmünder *et al.*, 2010) benefit from environmental life cycle assessment (LCA) literature covering systems of 10 kW or less. However, although small hydropower is an important rural electrification option proposed for its low environmental impact (Blanco, Secretan, and Mesquita, 2008; Greacen, 2004; Khennas *et al.*, 2000; Pereira, Freitas, and da Silva, 2010), these impacts have not been explored thoroughly for hydropower schemes smaller than 300 kW (Pascale, Urmee and Moore, 2011).

Life cycle assessment is a method of quantifying and understanding the environmental impacts of a product or service system from a whole of life perspective

(International Standards Organization, 1997) and has been used to explore the environmental impacts of various large scale electrification systems (IEA, 2002). A literature survey suggests that LCA has yet to be fully utilized to inform discussion and decision making around smaller scale rural electrification systems. The few studies applying LCA to rural electrification in developing countries thus far have focused on solar home systems (Alsema, 2000) and biodiesel (Gmünder *et al.*, 2010).

Rural electrification and the provision of low-cost, low-emission technology in developing countries require decision makers to be well informed on the costs, appropriateness and environmental credentials of all available options. While cost and appropriateness are often shaped by local considerations, environmental concerns are increasingly influenced by global concerns and are often harder to determine and convey to all stakeholders.

This chapter aims to quantify the potential environmental impacts of a community-sized hydropower system using LCA and compare the results with alternative electrification options for the community. Alternative options include a diesel generator and connection to the Thai power grid. In order to aid discussion on the appropriateness of LCA to rural electrification projects, this study also performs life cycle cost analysis on the hydropower system and the alternatives.

2.2 Life cycle assessment and life cycle costing

The International Organization for Standardization (ISO) provides a set of standards aimed at guiding LCA practice. ISO 14040 describes the basic LCA framework displayed in Figure 2.1. ISO 14040 also documents the principles of LCA. ISO 14044 provides additional information on aspects of LCA including the definition of goal and scope, life cycle inventory assessment, life cycle impact analysis, LCA interpretation and LCA reporting (ISO, 2010).

Conducting an LCA study using recognised standards in a transparent, diligent and fully documented manner aids LCA in meeting its intended application – its purpose – as well as in retaining credibility for continued modification, improvement and use.

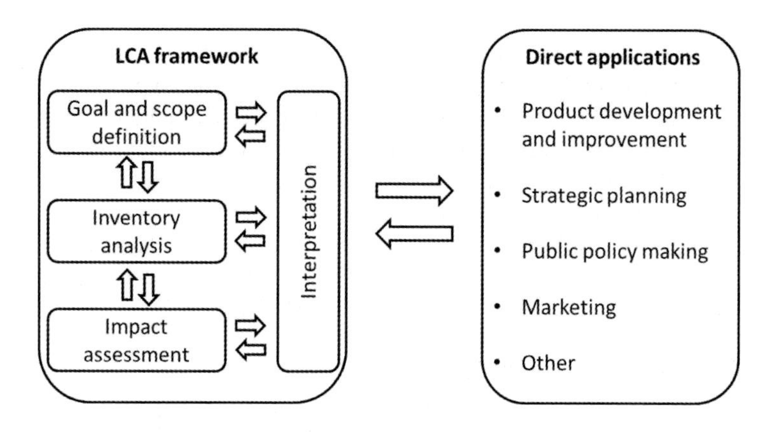

Figure 2.1 *LCA framework (ISO, 1997).*

LCA studies generally begin with defining the goal and scope of the study to ensure the vision, aims, methods and applications are clearly stated and understood by all participants. The scope defines the boundaries of the study in relation to geographic location, technology, time period and which environmental impacts will be considered (ISO, 1997).

With the goal and scope defined, the next stage of an LCA study is the data collection or life cycle inventory (LCI). In this stage, data are gathered for all considerations deemed within the LCA's scope and boundaries. LCI data detail all of the inputs and outputs involved with a product or service occurring within a specified boundary. Examples of common inputs and outputs include quantities of materials, energy, natural resources and pollutants (Guinee (ed) *et al.*, 2001).

Life cycle inventory analysis (LCIA) is the next stage in LCA and is the process of classifying, converting and aggregating LCI data to express a set of categorised impact results, for example, global warming potential. Each categorised impact is expressed using a reference unit, for example, kilograms of carbon dioxide equivalents (kg CO_2-e). Many classification and characterisation methods, such as CML 2001 and 2007, TRACI, EDIP, are recognised and can be used to convert inventory results into results for each impact category (GaBi 4, 2006). Characterisation methods are concerned with general broad areas of environmental protection, including 'human health, natural resources, the natural environment and the man-made environment' (Guinee *et al.*, 2001). Results for each impact category are presented in relation to a functional unit, for example, kilograms of greenhouse gas emissions per kilowatt hour of electricity supplied to the final consumer. LCIA results can optionally be packaged for communication and comprehension in a number of ways including normalisation and weighting (GaBi 4, 2006).

LCA is an iterative process, meaning that each stage of the process is revisited and revised many times as the study is conducted. Assumptions that are made during the study (due to data of time limitations) are clearly justified, documented and tested using sensitivity analysis to assist in the interpretation of the results.

LCA is not intended to be a decision-making process, but rather a transparent decision support tool to aid the decision-making processes. ISO 14040 points out that environmental LCA alone may not be appropriate in settings where the economic and social impacts of a service system are of central importance.

Life Cycle Costing (LCC) is particularly suited to transparently calculating a project's cost using a whole life approach including capital costs, operation, maintenance, transport and final disposal. When looking at rural electrification, LCC can be an important tool as the ongoing costs can be crucial in ensuring the long-term success of a project. A report in 2000 on micro-enterprise assisted by renewable energy stated that, 'although LCC can be used for any project, it is particularly appropriate for renewable energy (RE) projects of 5 to 10 years or more in which the fuel and repair costs of generators, for example, become considerable' (Allerdice and Rogers, 2000).

2.2.1 Methods and materials

As its core process, this study presents an LCA and an LCC case study of a community-sized hydroelectric system in rural Thailand. LCA reporting guidelines are drawn from ISO standards, best practice LCA examples (e.g. Moore, 2009) and from

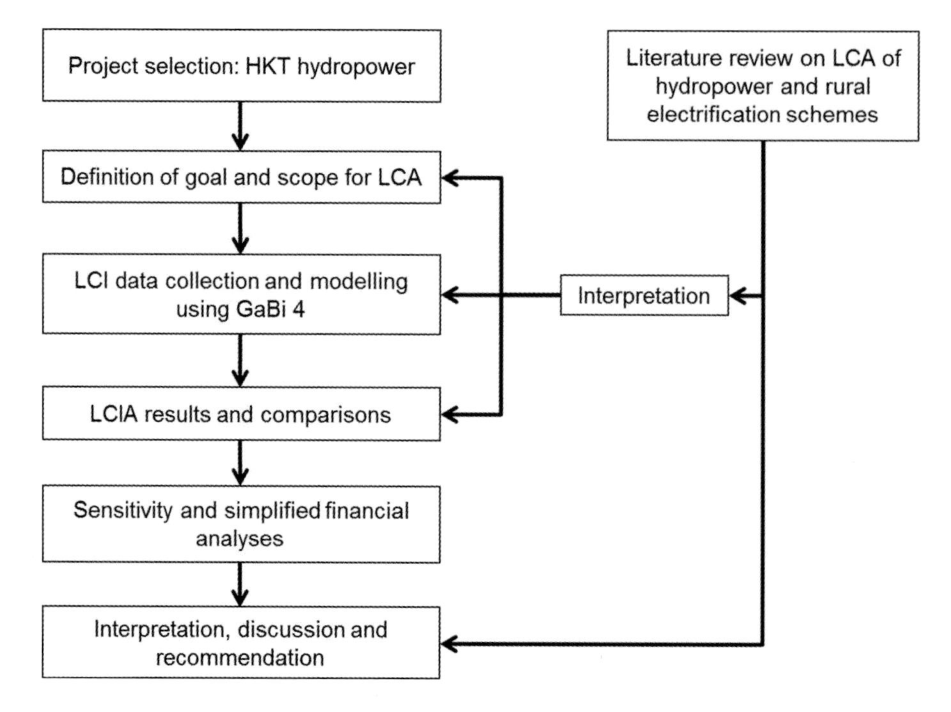

Figure 2.2 *LCA study on small hydropower scheme (Pascale et al., 2011).*

industry standard environmental declaration systems such as the Environmental Product Declaration® system. The LCA/LCC software GaBi 4 was chosen due to the availability of the software and expert advice (GaBi 4, 2006).

LCA modelling and impact assessment on the project followed an iterative process:

- Life cycle inventory data collection on a specific community hydropower scheme and the alternative power supply options
- Data and process entry into the GaBi 4 educational LCA software and
- LCIA and comparison of alternative options.

After the presentation of the final iteration's LCIA result, a sensitivity analysis and LCC analysis are performed to further inform results and identify the important areas for further research. Figure 2.2 details this study's approach to the hydropower scheme LCA.

2.2.2 Project selection

A 3-kW community hydropower system located at Huai Kra Thing (HKT) was chosen for the study because of the good level of data availability from the Border Green Energy Team (BGET) that had constructed the project. The main author of this chapter played a central role in the project management team.

HKT village is located in Tak Province in rural Thailand. Tak Province is one of 76 provinces in Thailand and features large areas of rugged mountainous terrain.

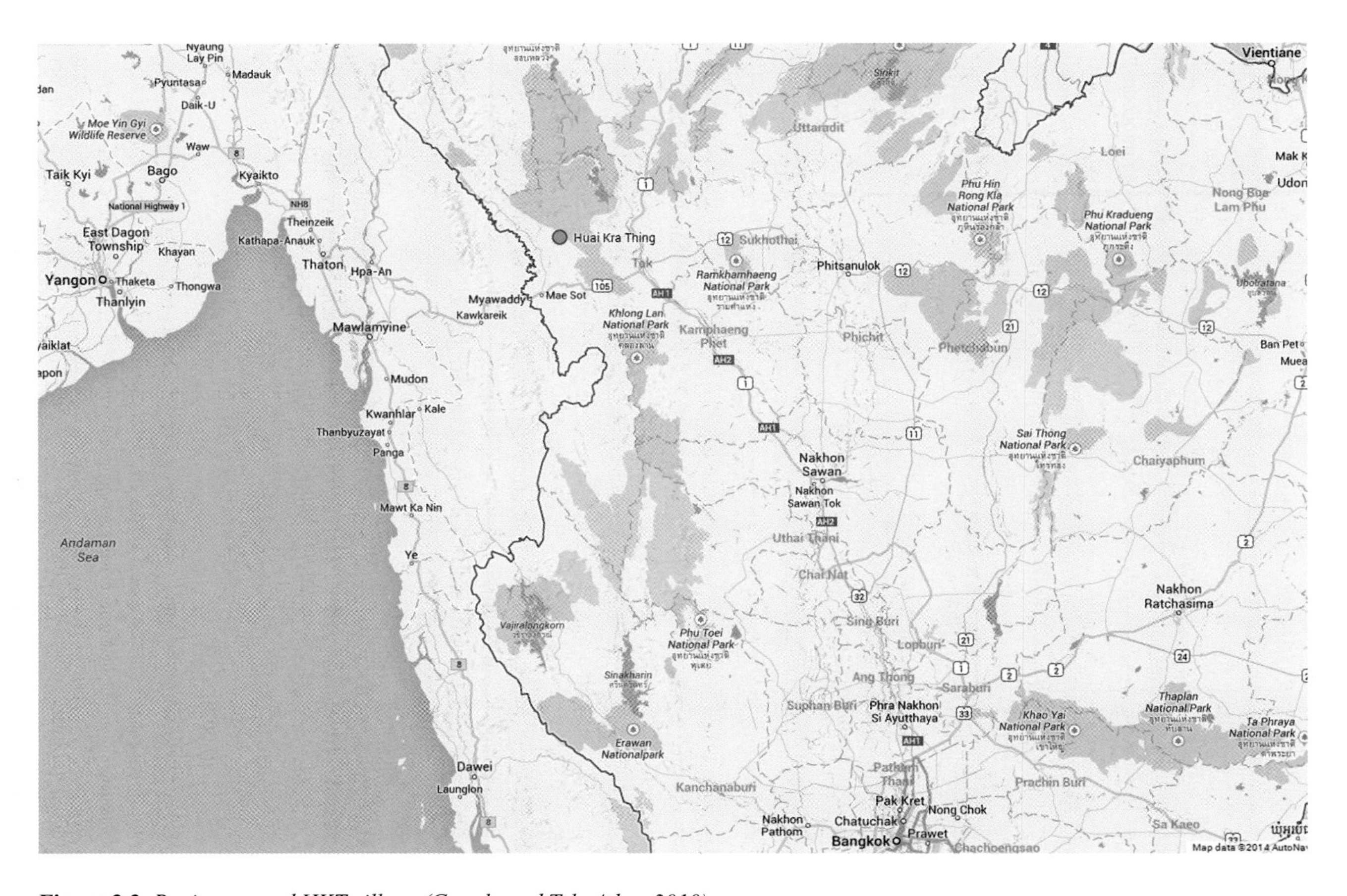

Figure 2.3. *Region around HKT village (Google and Tele Atlas, 2010).*

It also shares a border with Burma's Kayin State (Kawthoolei). While HKT can be reached from the capital of Tak, the Thailand/Burma border town of Mae Sot is of more importance as a local hub for HKT residents and local aid organisations.

HKT village is, by road, approximately 61 kilometres from Mae Sot. The village centre consists of approximately 45 houses and has a population of 230, the majority of whom are ethnically a Thai-Karen mix. A charismatic and ritually tattooed village elder (with a penchant for Hawaiian shirts and cigars!) serves as the village headman and de facto judge, foreman and statesman. Agriculture is the main occupation of the villagers, although entrepreneurs offer a variety of services and goods (KNCE, 2007). Villagers rely on animal husbandry and the surrounding forest to meet their remaining subsistence needs.

The village road system consists of a single central road allowing access to important community buildings. While the road is mainly constructed using dirt, areas prone to mud slides and bogging in the rainy season are paved over with concrete, which means access by heavy equipment is extremely limited. Most villagers own or have access to a motorbike or bicycle and a few own personal vehicles such as 4WD trucks. Almost all personal transport in the village takes place on foot. Community structures in the village include a Baptist church, a Catholic church, a primary school, school teacher's quarters, a medical clinic, an open air community meeting centre and water storage tanks.

Water is available from a gravity feed water system plumbed to most household taps. Water enters the gravity feed system roughly 700 metres from the village's central water storage tanks and flows in polyvinyl chloride (PVC) pipe from intake to tanks. Water is then distributed from the tanks to village houses via a network of PVC pipes.

Wood fires serve as the primary energy source for cooking and lighting for evening socialising. Cooking fires are built in sand pits at the centre of well-ventilated kitchens. Open wood and charcoal fires at ground level provide the heat needed to smoke and roast meat as well as distil local rice whiskey. Direct solar radiation is used to dry village staples such as rice, chillies, bananas and fish. Lighting from fires has been supplemented over the years by candles and electricity supplied by small diesel generators, a wide array of batteries and government-supplied solar home systems (Allerdice and Rogers, 2000; Lynch *et al*, 2006).

The HKT village hydroelectric system was constructed in early 2006 through the combined efforts of Thailand/Burma border organisations, local villagers, refugee camp students, students from the United States and the United Nations Development Programme (Greacen, 2006). The HKT village hydroelectric system harnesses energy from a nearby river to supply all-year-round electricity to seven of the community buildings.

2.2.3 Scope

This study models the HKT village hydroelectric system over a 20-year period to enable comparison and discussion in relation to rural electrification alternatives such a solar PV and wind (Alsema, 2000; Kenfack *et al.*, 2009; Varun and Prakash, 2009). The lifespan of the hydroelectric system components was estimated using observations of installation methods, a literature survey, anecdotal reports from technicians employed by BGET and expert judgement. It was assumed that no components would last longer than 20 years.

Scheme components sourced from local organic renewable resources were assumed to be replaced every two to five years. However, while locally sourced wood and bamboo used for construction purposes were included in the model, impacts from their use were not included. Local organic renewable resources were treated as having a zero net impact cycle, where they:

- Did not leave their regional ecosystem
- Were not treated with a preservative
- Were harvested and transported using human power and
- Were allowed to decompose *in situ* after use

It was assumed that emissions associated with decomposing organic renewable resources such as wood and bamboo were biogenic and therefore were not included in calculation of the global warming potential of the system.

The sole function of the HKT hydroelectric system considered in this chapter was generation of electricity so that all the environmental burdens were allocated to the electricity produced. Neither the co-production of additional forms of energy (such as the use of dump load heat energy to do work) nor the potential use of any section of the hydropower scheme for irrigation, aquaculture or recreation was considered in this study.

This LCA models the HKT scheme as installed in early 2006. While the model covers the provision of an electrification infrastructure in HKT village, house wiring and core energy services were excluded from the study as they were required, regardless of whether the energy was supplied by micro hydro, diesel generator or from the grid.

It has been stated that the quality and annual availability of power produced by an electricity generating system has not always been adequately addressed in LCAs (Gagnon *et al.*, 2002). As a means to adjust for intermittency or annual availability issues, Gagnon *et al.* (2002) advises that the LCA of an intermittent or seasonally fluctuating scheme should include all backup power sources required to maintain a minimum level of service. The HKT installation was designed to provide year-round, uninterrupted power for community buildings, and therefore a backup power system was not included in our model. The HKT hydropower system was modelled to operate at 53.3 per cent of the rated capacity[1] for 85 per cent[2] of the year.

2.2.4 Functional unit, energy availability and system losses

A common functional unit in LCA studies on electrical power generation is one kilowatt-hour (1 kWh) of electrical energy; however, studies vary in where that 1 kWh is measured and whether the functional unit includes consideration of system losses (IEA, 2002).

[1]The turbine generation set point after installation was 1.6 kW. This is 53.3 per cent of 3 kW turbine's rated capacity.

[2]Anecdotal reports from area local people suggest considerable downtime for HKT and other regional similarly sized and constructed hydropower installations.

The functional unit used in this study represents 1 kWh of electrical energy available for consumption by the villagers at the point of connection to the community structures. The functional unit did not concern itself with how energy was used by villagers, only that the energy was available to them. All system losses incurred after generation by the turbine (transmission, distribution and electrical conditioning) were included in the calculation of the functional unit (Pascale *et al.*, 2011).

2.2.5 Human labour and maintenance

Human labour, which accounts for all the energy used during system installation, was outside the scope of this LCA. Likewise, while materials and transportation for equipment replacement were included in the model, human labour needed for maintenance and replacement was excluded.

2.2.6 Packaging, paints and lubricants

Packaging was used to facilitate the sale and safe transport of equipment from purchase to installation. However, this LCA does not include packaging material in the modelling of system components. Packaging materials were estimated to be less than 1 per cent of the total system environmental impacts and were excluded. Paints and lubricants used on system equipment such as the turbine were also expected to be less than 1 per cent of the total system environmental impacts (for the selected environmental impact categories) and were excluded.

2.2.7 Disposal, recycling and re-use

When equipment has reached the end of its functional life, it is generally stockpiled (locally) until another use is found for it or villagers might gain some benefit (e.g. a recycling scheme) from disposing of it regionally. Given this context, neither was reuse of system materials nor recycling included in this LCA model.

2.2.8 Water consumption

While a UNESCO-IHE Institute for Water Education report associates water use from hydropower mainly with water evaporation and seepage from large reservoirs, the OECD states explicitly that, 'Water used for hydroelectricity generation is an in situ use and is excluded' (Gerbens-Leenes *et al.*, 2008; OECD, 2009). Due to the small reservoir size and the return of water used for power generation to the river, water flowing through the system during operation was not tracked in this study. However, water consumed in the production of all scheme equipment and materials was included in this LCA.

2.2.9 Land use

Land use was tracked in this LCA. However, aside from assigning a generic land use change category, further land use change differentiation was not pursued and not fully included in the formal LCA results. A lack of clarity around appropriate inclusion of land use in software modelling also appears to have hindered the formal inclusion of land use in recent hydropower and rural electrification LCAs (Gmünder *et al.*, 2010; EPD, 2005).

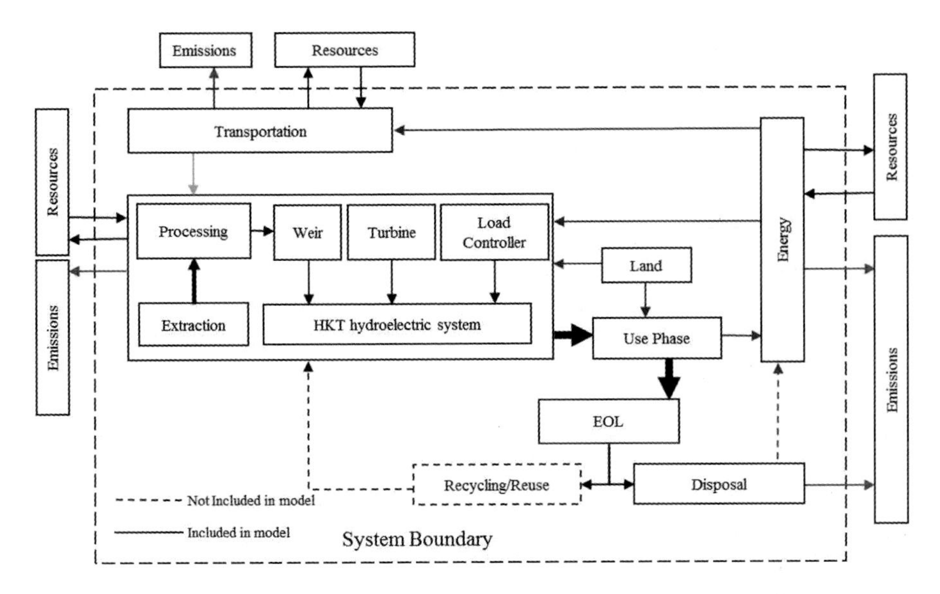

Figure 2.4 *Simplified system process drawing showing scope and system boundaries [21].*

2.2.10 System boundaries

The system boundaries of the study were from 'cradle to grave' and include the extraction and processing of resources to make products and materials comprising the HKT hydropower system over its life cycle. Resource extraction and processing, materials, product manufacturing, construction, operation and equipment were then followed through the use phase to their end-of-life phase. Transportation of all equipment from point of manufacture to HKT was included in the model. Figure 2.4 shows the processes and system boundaries included.

2.3 Life cycle inventory exclusion cut-off criteria

The HKT hydroelectric system comprised multiple product systems made up of many unit processes. ISO 14044 (2006) recognises that LCA is an iterative process and that inclusion of all inputs, outputs and processes in an LCA requires a substantial time commitment and is not practical. Cut-off criteria provide an approximate guide for the thoroughness of the LCA and detail not only what was included in the LCA, but also equally as important, estimate what was not included (Guinee *et al.*, 2001).

This study included at least 95 per cent of the total mass and total energy inputs of the life cycle of the HKT hydropower scheme. All individual mass flows comprising more than 1 per cent of the total mass flows were included. All individual energy flows above 2 per cent of the total energy input were included. If an individual mass or energy input flow is below the cut-off criteria, but was expected to have environmental relevance, it was included in the study. Cut-off criteria for this study were based on ISO 14044 suggestions for mass, energy and environmental relevance (ISO, 2010).

2.3.1 Data collection and process selection

Modelling of the hydropower system, diesel system, grid connection and transportation modes and distances draws on hydropower system documentation, industry literature and project management estimations to specify life cycle inventory materials and processes. LCI data were then matched with the available GaBi 4 database materials and manufacturing processes. Although Thailand's power grid mix was the only specifically Thai GaBi 4 dataset currently available, lack of Thailand-specific LCI data was not expected to be a major issue because of a strong commitment to international industrial standards, and the GaBi 4 database represents industry standard processes (TISI, 2010). It was recommended, however, that future studies should use specific Thai life cycle databases as they become available.

2.3.2 Impact categories and indicators

Table 1 lists the base set of CML 2001 impact categories and indicators reported in this study. The primary energy demand was the total amount of energy consumed by the scheme over its 20 year life span. Life Cycle Impact Assessment (LCIA) results were scaled to the lifetime energy production in kWh and presented in a reference unit for each LCIA category.

Table 1 *LCA categories addressed in this study (Pascale et al., 2011).*

Life cycle impact assessment categories and indicators	*Unit / kWh*	*Short form*	*Description [17]*
CML 2001 – Dec. 07, Global Warming Potential (GWP) (100 years)	kg CO_2-e	GWP	emissions contributing to climate change
CML 2001 – Dec. 07, Acidification Potential (AD)	kg SO_2-e	AP	Emissions causing acidification of the environment
CML 2001 – Dec. 07, Eutrophication Potential (EP)	kg PO_4-e	EP	emissions increasing environmental nutrient levels
CML 2001 – Dec. 07, Ozone Layer Depletion Potential (steady state) (ODP)	kg R11-e	ODP	emissions causing thinning of the stratospheric ozone layer
CML 2001 – Dec. 07, Photochemical Ozone Creation Potential POCP)	kg ethane-e	POCP	airborne pollutants creating ozone through interaction with sunlight
CML 2001 – Dec. 07, Abiotic Depletion (ADP)	kg Sb-e	ADP	depletion of non-living resources
Primary energy demand from renewable and non-renewable resources (net cal. value) (PED)	kWh	PED	total consumption of resources (GaBi 4, 2006a)

2.4 LCC considerations

Life cycle costing for the HKT hydropower scheme uses financial information gathered from budgets prepared by project partners for UNDP grant reporting. Estimations of ongoing costs were based on discussion with local project staff. The analysis period was chosen to match the 20-year focus of the LCA.

Costs reported in this study were those at February 2006 in Thai Bhat (THB) as that was the local currency for this study. At that time 40 Thai Bhat equalled 1.02 USD, 0.85 EUR and 1.36 AUD (XE.COM, 2013). The calculations used an inflation rate of 3.22 per cent, which was the average Thai inflation rate over 2005 to 2009 (Bureau of Trade and Economic Indices, 2010). As suggested by Allderdice and Rogers (2000), this study used a discount rate of 10 per cent.[3]

The LCC analysis included costs associated with installation, operation, maintenance, diesel fuel, electric utility rates and major component replacements. The LCC does not include the sale of electricity from the hydropower or diesel electrification schemes to customers, subsidies, taxes, externality costs, minor component replacement, costs involved with alternative financing mechanisms such as loans or village labour. For the same reasons, the LCA does not include re-use of system materials or recycling.

2.5 Model descriptions

The following section provides a description of the most relevant features of each system used in this model.

2.5.1 Hydropower scheme overview

The Huai Kra Thing hydroelectric power system, installed early in 2006, was a 3 kW system run from the river system.[4] The system was designed for all-year-round electrification of seven community-oriented buildings. System design followed a 'soft approach' which minimised large permanent structures and utilised locally sourced and easily replaceable structures (Greacen, 2004). Figure 2.5 provides a basic overview of the community hydroelectric system and has separated the scheme into seven distinct physical sections: weir, intake, water canal and forebay, penstock, powerhouse,

[3]Arguments might be made for other rates. If approaching the HKT scheme purely as a project competing for investment against hydropower projects of a similar size and type (electrification mainly for lighting) from around the world, the average IRR for such projects was 7.75 per cent (Khennas *et al.*, 2000) If the HKT scheme was viewed purely as a rural microenterprise project, a discount rate of 50 per cent was suggested (Varun *et al.*, 2009). If used to compare results directly with a relatively recent LCC of a solar home system in Thailand, a discount of 5 per cent should be used.

[4]The installation is not described in this section but a brief description can be found in Greacen (2006). More in-depth but generic description of community hydropower installations can be found in physical book form (Inversin, 1986; Maher and Smith, 2001).

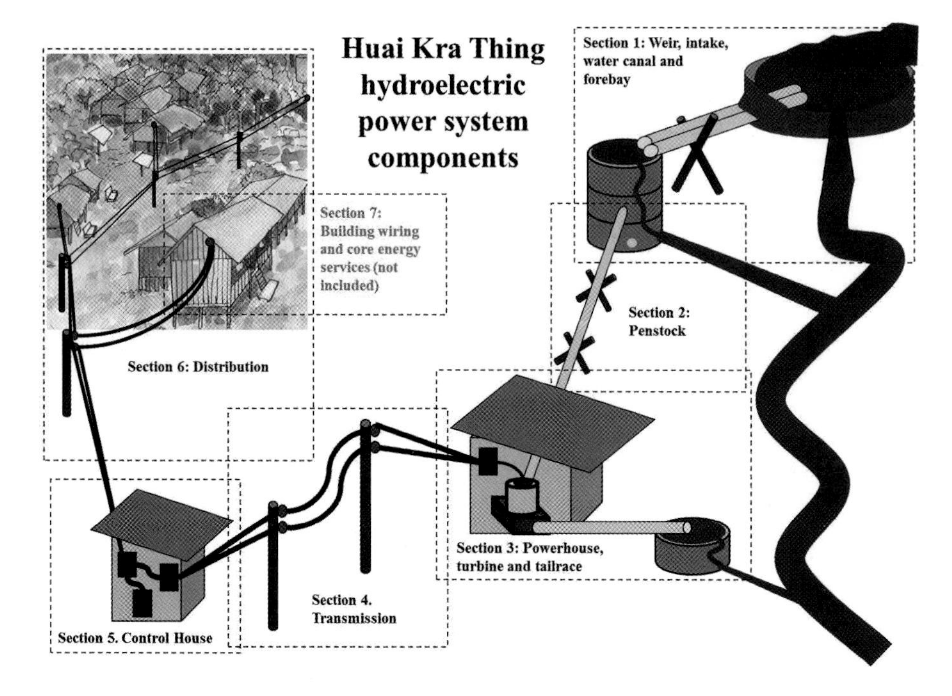

Figure 2.5 *Overview of the HKT community hydroelectric system separated into physical sections. Karen village painting (Pascale et al., 2011; Sein, 2006). Table 2.2 summarises the major system inputs and outputs over the 20-year period.*

Table 2.2 *Significant hydropower system inputs and outputs (includes replacements over 20 years).*

Item	*Description*	*Quantity*	*System sections*
A	Dry cement mix	2098.7 kg	1, 2, 3, 4, 6
B	Polyvinyl chloride (PVC) pipe parts	694.0 kg	1, 2, 3
C	Aluminium wire with PVC sheathing	223.0 kg	2, 3
D	3kW pump as turbine unit (Pump and motor)	480.0 kg	3
E	Cast Aluminium materials	66.7 kg	3, 4, 5, 6
F	Ceramic materials	69.9 kg	4, 5, 6
G	Galvanised steel materials	102.7 kg	1, 2, 3, 4, 5, 6
H	Hand wound 380V to 230V power transformer	80.0 kg	5
I	Motor-run capacitors	20.0 kg	3
J	3kW Induction generator controller (IGC)	3.8 kg	5
K	Forest use change	94 m^2	1, 2, 3, 4, 5, 6
L	Potential riverbed use change	470 m^2	1, 2, 3
M	Lifetime energy available at connection to buildings	218,977 kWh	7
N	Maximum waste heat to water	258,404 kWh	1, 2, 3

turbine and tail race, transmission, control house, distribution and building wiring and core energy services. Core energy services are not included.

2.5.2 Weir, intake, canal and settling tank

The goal of this part of the hydroelectric system (Figure 2.6) was to provide a clean and steady flow of water to the penstock. Each individual component in this section had a specific role in achieving that goal. The weir's function was to resist the flow of water in the stream enough to keep the intake to the canal submerged while allowing floating debris to pass over the structure. Located just behind the weir, the intake allowed water to enter the canal and used a filter to limit the size of debris entering the canal. The canal transported water to the settling tank. The settling tank aimed to keep the penstock full while allowing any debris in the water to settle to the bottom of the tank.

Over 91 per cent of this section by mass used local resources consisting mostly of sand, rocks, gravel and wood. The remaining 9 per cent included materials such as polyvinyl chloride (PVC) pipe and dry cement mix, which due to life spans of 20 years or greater as suggested by industry literature, produced this section's most significant environmental impacts. The small reservoir (approximately 25 m^2) created behind the weir was expected to have a minimal environmental impact. There was a potential for riverbed exposure and significant environmental impact from the removal of water

Figure 2.6 *Photographs of the weir and intake (upper left), the settling tank (upper right) and the canal as installed in HKT.*

Table 2.3 *20-year life cycle materials' input for the weir, intake, canal and settling tank.*

Item	*Quantity*	*Unit*	*HKT Total Mass (kg)*
Bags of dry cement mix for intake	0.67	bags	33.33
Bags of dry cement mix for settling tank	2.00	bags	100.00
Schedule 40 - 4" inch blue PVC pipe for canal	50.00	m	149.56
Schedule 40 - 1" blue PVC pipe cleanout valve	1.00	pce	0.05
Concrete rings for settling tank (made off site-purchased whole)	3.00	pce	150.00
Galvanized nails for fastening canal supports	60.00	pce	0.70
HDPE tarp used as stream bed liner at weir	8.00	sqm	6.22
PVC glue/solvent for canal	1.00	can	1.00
Bailing wire to tie plastic mesh filter to bamboo structure	12.00	m	0.07
Bailing wire to tie mesh filter and to tie canal to stands	92.00	m	0.56
50 kg woven PP rice bags (purchased not reused) to hold rocks and sand for weir	133.33	pce	0.01
Plastic mesh filter (1cm^2) used to filter debris at intake	1.33	sqm	0.00
Plastic mesh filter (1cm^2) used to filter debris in settling tank	1.33	sqm	0.00
Local wood support structure for canal	20.00	pce	1585.53
Gravel Mixers added to dry cement to make wet concrete for weir/intake	64.34	kg	64.34
Locally collected rocks for weir	1666.67	kg	1666.67
Gravel Mixers added to dry cement to make wet concrete for settling tank	193.02	kg	193.02
Sand Mixers added to dry cement to make wet concrete for weir/intake	33.86	kg	33.86
Locally collected sand for weir	1666.67	kg	1666.67
Sand Mixers added to dry cement to make wet concrete for settling tank	101.59	kg	101.59
Water and Air Mixers added to dry cement to make wet concrete for weir/intake	5.30	kg	5.30
Water and Air Mixers added to dry cement to make wet concrete for settling tank	15.89	kg	15.89
Bamboo twist ties for weir	133.33	pce	0.67
Woven bamboo filter structure for intake	5.00	kg	5.00
Woven bamboo filter structure for settling tank	5.00	kg	5.00
Forest and river to weir and reservoir	25.00	sqm	
Forest to canal	6.35	sqm	
Forest to settling tank	0.50	sqm	
Reduced flow in river between intake and settling tank	70.00	sqm	

from the river during this section's used phase process.[5] Table 2.3 shows the material and land inputs used in modelling this section of the system.

2.5.3 Penstock

The penstock transported water from the settling tank to the turbine. The HKT penstock covered a horizontal distance of 172 m and moved water through an approximate 30 meter change in altitude (head). The penstock consisted of a four inch (102 mm) PVC pipe connected using a PVC solvent. The penstock was stabilised where necessary using concrete and local wood. Water was pressurised over the entire length of the penstock. Figure 2.7 shows the main components of this section.

Environmentally significant materials in this section were PVC pipe and dry cement mix, both of which were, given industry literature, suggested life spans of 20 years or greater. There were no significant used phase processes for this section of the hydropower scheme. Table 2.4 shows the 20-year life cycle materials' inputs and land inputs used in modelling of this section of the system.

Figure 2.7 *Photos of penstock as installed in HKT.*

[5]Although HKT system design stipulates that no more than 50 per cent of the stream's water should be used for power generation, in order to constantly reach that goal, more than half of the water may need to be removed from the river between weir and settling tank.

Table 2.4 *Twenty-year life cycle materials' inputs for the penstock.*

	Quantity	*Unit*	*HKT Total Mass (kg)*
Bags of dry cement mix for thrust blocks	3.00	bags	150.00
Schedule 40–4" inch blue PVC for penstock	172.00	m	514.49
Galvanized nails for fastening penstock supports	180.00	pce	2.10
PVC qlue/solvent for penstock	4.00	can	4.00
Bailing wire to tie penstock to stands	240.00	m	1.47
Local wood support structure for penstock	60.00	pce	4756.58
Gravel Mixers added to dry cement to make wet concrete for penstock	289.52	kg	289.52
Sand Mixers added to dry cement to make wet concrete for penstock	152.38	kg	152.38
Water and Air Mixers added to dry cement to make wet concrete for penstock	23.84	kg	23.84
Forest to penstock	21.84	sqm	
Reduced flow in river between setting tank and outflow	400.00	sqm	

2.5.4 Powerhouse, turbine and outflow

This part of the hydroelectric system harnessed the energy in the flowing pressurised water to generate electricity. The turbine, which took in pressurised water and output electricity and used water, made this possible. Huai Kra Thing's turbine consisted of an Ebara 4 kW centrifugal pump, which was being used in reverse as a 3 kW 'pump as turbine'[6] to generate electricity. A powerhouse building constructed mainly of local materials protected the turbine and associated electrical equipment. The spent water was released through the outflow, which consisted of a length of PVC pipe and a concrete ring that dispersed the water to minimise erosion. Figure 2.8 shows the main components of this section.

The dry cement mix and the turbine were the most environmentally significant materials in this section of the system. The life spans for components involving dry cement mix were assumed to be 20 years. Consultation with local experts suggested a five-year operational life for the turbine run in HKT conditions. In light of its initial cost and need for three more replacements in the 20-year period, the turbine was the most significant LCC item in the entire system. Table 2.5 shows the life cycle material and inputs used in modelling this section of the system.

Used water re-joined the stream approximately 7 m after it exited the outflow pipe. At 10 litres a second and with the system operational 85 per cent of the year,

[6]Reference material on PATs can be found in (Greacen and Kerins; Williams, 1995).

Figure 2.8 *Photos of the powerhouse, turbine and outflow from HKT.*

Table 2.5 *Twenty-year life cycle material inputs for the powerhouse, turbine and outflow.*

Item	*Quantity*	*Unit*	*HKT Total Mass (kg)*
Galvanized steel sheets for powerhouse roofing	6.25	sqm	36.91
Galvanized nails for fastening powerhouse structure	628.00	pce	10.82
Bags of dry cement mix for concrete pad for turbine	4.00	bags	200.00
Schedule 40 – 6" inch blue PVC pipe for outflow	3.00	m	24.69
Schedule 40 – 4" to 6" male PVC adapter for outflow	1.00	pce	1.00
Schedule 40 – 4" inch blue PVC pipe screw adapter for outflow	1.00	pce	0.50
Schedule 40 – 4" inch blue PVC pipe 45 degree connector for buildup	2.00	pce	0.88
Schedule 40 – blue PVC pipe 4" to 3" reducer for buildup	1.00	pce	0.84
Schedule 40 – 3" inch blue PVC pipe screw adapter for buildup	5.00	pce	2.00

(*Continued*)

Table 2.5 *Continued*

Item	*Quantity*	*Unit*	*HKT Total Mass (kg)*
Concrete rings for outflow (made off site-purchased whole)	1.00	pce	50.00
25uF Motor run capacitors (380V)	40.00	pce	10.00
50uF Motor run capacitors (380V)	40.00	pce	10.00
Plasting housing for motor run capacitors and overspeed circuit breaker	1.00	pce	1.00
Bailing wire to tie down penstock in powerhouse	200.00	m	1.23
3" bronze gate valve for turbine/penstock buildup	1.00	pce	2.85
3" cast iron connector for turbine/penstock buildup	1.00	pce	1.07
7/16 steel J bolts/nuts/washers for mounting turbine to foundation	4.00	set	0.80
Single throw two pole breaker (6A) in power house for overspeed	1.33	pce	0.26
Red LED system operating light in powerhouse	4.00	pce	0.0014
Local wood for powerhouse structure	188.11	m	1590.66
Gravel Mixers added to dry cement to make wet concrete for power house foundation	386.03	kg	386.03
Sand Mixers added to dry cement to make wet concrete for power house foundation	203.17	kg	203.17
Water and Air Mixers added to dry cement to make wet concrete for power house foundation	31.78	kg	31.78
Bamboo walls for powerhouse structure	70.54	kg	70.54
Bamboo form and rebar for concrete pad	4.00	kg	4.00
4 kW Ebara end suction volute pump with four pole 380 V motor	4	pce	480.00
Forest to outflow	1.11	sqm	

5,364,792 m^3 of water would have been used over the 20 years of power generation. Table 2.6 shows the details of the calculation of total lifetime water use for the power generation. As discussed earlier, water used for power generation was not considered as being consumed in the LCA model.

2.5.5 Transmission

The function of this section of the hydroelectric system was to transmit electricity safely back to the control house where it was regulated and conditioned for distribution to the community. Electricity was transmitted along 600 meters (one-way) of 25 mm^2 aluminium cable. The transmission wire was raised off of the ground by 3 m tall, locally sourced wooden poles and connected to those poles using aluminium-framed ceramic insulators. Wooden power poles were placed in hand-dug holes and then concrete added to the hole to provide increased stability (Figure 2.9).

Table 2.6 *Calculation of water use for the HKT turbine.*

a. Water flow	10	l/s
b. Seconds in an hour	3600	seconds
c. Hours a day	24	hours
d. Days a year	365.25	days
e. % year system operational	85%	
g. Years of system study	20	years
f. Annual water used by turbine = a * c * d * e =	**268,240**	**m³**
h. Lifetime water use calcualtion = f * g =	**5,364,792**	**m³**

Figure 2.9 *HKT transmission poles.*

Dry cement mix and aluminium wire were the most environmentally significant materials in this section. Industry literature suggested life spans for these materials of 20 years or greater. This section contains no significant use phase processes. Table 2.7 shows the life cycle material and land inputs used in modelling this section of the system.

2.5.6 Control house, transformer, dump load and induction generator controller (IGC)

This section controls turbine power output, prepares transmitted electricity for distribution to the village and sheds excess energy that was generated by the turbine but not consumed by the village. Central components of the control and conditioning equipment included a 3 kW transformer, which stepped electricity down from 380 V to 230 V and a 3 kW induction generator controller (IGC) that kept a constant load on the turbine and diverted excess power to a ballast load. The 3 kW ballast load was located in the control house. A small control house building protected the control equipment from damage and also protected the villagers and roaming animals from harm. Figure 2.10 shows the components from this section.

The transformer and the IGC were the most environmentally significant materials for this section of the system and were replaced every five years. Such a replacement schedule is in line with the life spans of the electronic components as given in other rural electrification literature (Kenfack *et al.*, 2009). There were no significant use phase processes for this section of the hydropower scheme. Table 2.8 lists the life cycle material and land inputs used in modelling this section of the system.

Table 2.7 *Twenty-year life cycle material inputs for the transmission system.*

Item	*Quantity*	*Unit*	*HKT Total Mass (kg)*
Bags of dry cement mix for transmission power pole bases	16.00	bags	800.00
Aluminum Wire 25sqmm for transmission - sheathed	1,200.00	m	136.20
Ceramic insulators in aluminium bracket for transmission line	40.00	pce	62.00
Aluminum wire to provide stress relief for transmission wires at pole - no sheath	80.00	m	0.68
Galvanized nails for fastening insulator structure to transmission pole	320.00	pce	3.73
Local wood for power transmission poles	160.00	pce	9513.15
Gravel Mixers added to dry cement to make wet concrete for transmission line	1544.13	kg	1544.13
Sand Mixers added to dry cement to make wet concrete for transmission line	812.70	kg	812.70
Water and Air Mixers added to dry cement to make wet concrete for transmission line	127.12	kg	127.12
Forest to transmission	15.39	sqm	

Figure 2.10 *Control house and power conditioning and control equipment from HKT.*

Table 2.8 *Twenty-year life cycle material inputs for the control house, transformer, ballast load and IGC.*

Item	*Quantity*	*Unit*	*HKT Total Mass (kg)*
Ceramic holders for resistive heating elements	6.00	pce	0.30
Galvanized steel sheets for control house roofing	6.25	sqm	36.91
Galvanized nails for fastening control house structure	608.00	pce	7.09
380 to 230Vac 50Hz Hand Wound Power Transformer	4.00	pce	80.00
Steel ballast box (.25 x .8m locally made)	1.00	pce	8.00
Stainless steel electronics box - manufactured regionally	1.00	pce	5.00
23inch long 1kW resistive elements for ballast loads (FeCrAl)	6.00	pce	3.00
Single throw two pole breakers (15A and 20A) in control house	2.67	pce	0.53
Red LED system operating light in control house	4.00	pce	0.0014
Local wood for control house structure	183.15	m	1548.78
Bamboo walls for control house structure	68.68	kg	68.68
3kW Electric Load Controller (IGC) from Nepal	4	pce	3.8075
Forest to control house building	4	sqm	

2.5.7 Distribution

This section of the HKT hydroelectric power scheme used 1178 m of 16 mm^2 aluminium wire (insulated with PVC) to distribute the electricity from the control house to seven community buildings. The community-oriented buildings included in the distribution network were a Baptist church, a Catholic church, the village headman's house, a medical clinic, a primary school, the school teacher's quarters, and an open air community meeting centre (Figure 2.11).

Dry cement mix and aluminium wire were the most environmentally significant materials in this section. Industry literature suggests life spans for these materials to be 20 years or greater. The availability of electrical power at point of connection to community buildings was the central use phase process for this section. Even with estimated losses in transmission (1.52 per cent), conversion (5.95 per cent) and distribution (0.85 per cent average), 1.47 kW of the 1.60 kW of power (in total 92 per cent) produced by the turbine was expected to be available at points of connection to community buildings. Table 2.9 lists the life cycle material and land inputs used in modelling this section of the system.

2.5.8 Scaling impacts to function unit

All LCIA results produced by the GaBi 4 model of the HKT hydropower system were scaled to the functional unit. This was achieved by calculating the total lifetime energy available to villagers at the point of entry to the community buildings and then dividing the LCIA results by the total lifetime energy. Total lifetime energy available to the villagers at the point of entry to community buildings was calculated to be 218,977 kWh as shown below:

Total lifetime energy = (average power available to community buildings) × (hours in a day) × (days in a year) × (per cent of year the system is operational) × (length of the study)

= 1.469 kW × 24 hours × 365.25 days × 85 per cent × 20 years
= 218,977 kWh

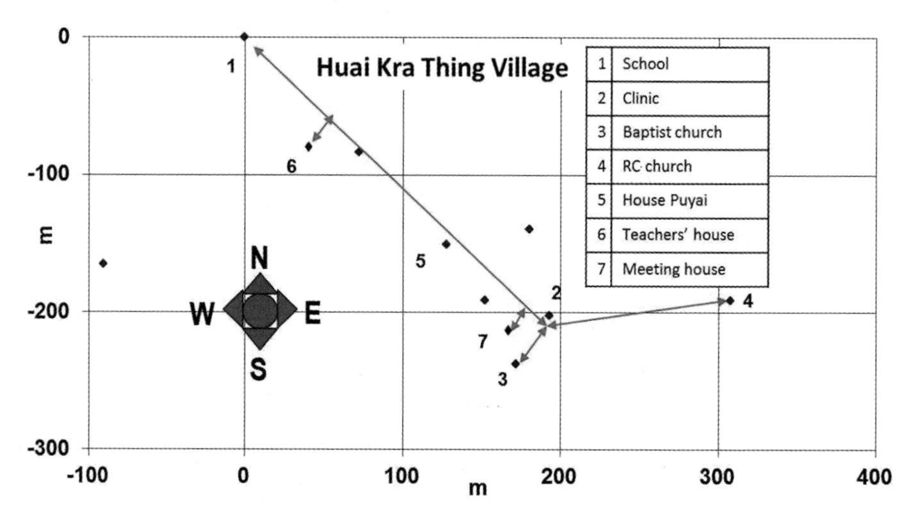

Figure 2.11 *Power distribution map for the HKT (modified from BGET, 2006).*

Table 2.9 *Twenty-year life cycle material inputs for distribution.*

Item	*Quantity*	*Unit*	*HKT Total Mass (kg)*
Bags of dry cement mix for distribution power pole bases	16.00	bags	800.00
Aluminum Wire 16sqmm for distribution - sheathed	1,178.00	m	86.82
Ceramic insulators in aluminium bracket for distribution line	47.00	pce	72.85
Aluminum wire to provide stress relief for distribution wires at pole - no sheath	94.00	m	0.80
Galvanized nails for fastening insulator structure to distribution pole	376.00	pce	4.39
Local wood for power distribution poles	160.00	pce	9513.15
Gravel Mixers added to dry cement to make wet concrete for distribution lines	1544.13	kg	1544.13
Sand Mixers added to dry cement to make wet concrete for distribution lines	812.70	kg	812.70
Water and Air Mixers added to dry cement to make wet concrete for distribution lines	127.12	kg	127.12
Forest to distribution	15.39	sqm	

2.6 Project costs and LCC calculations

Installation costs for the HKT hydropower system were calculated using finalised budgets prepared for the UNDP by a local project partner after the installation of the system (Pascale *et al.*, 2011). As this installation was part of a UNDP grant, installation costs include equipment costs, administration, coordination, capacity building, evaluation and outreach costs (KNCE, 2007). Table 2.10 lists the installation costs for the HKT hydropower system.

Local project partners also provided an estimate of the support (in kind) that was provided by all associated stakeholders to operationalise the project. In kind, support for the HKT hydropower scheme was estimated to be THB 551,875. The total installation cost for the project is THB 973,587.

Operation and maintenance costs for the HKT hydropower scheme consist of an estimated yearly stipend for the system caretaker to cover minor system maintenance. A THB 6,362 annual stipend for the system caretaker was calculated using Allderdice and Rogers's (2000) approximation of 3 per cent of capital costs, which for this study used only the initial THB 212,058 equipment costs. Major component replacement costs were estimated at THB 33,000 every five years and included transportation. Table 2.11 lists the operation and maintenance costs for the HKT hydropower system.

Table 2.10 *HKT hydropower system installation costs.*

Description	*Cost (THB)*
Honorarium (local project managers)	90,000
Coordination (telephone, document preparation)	14,965
Equipment (transportation, food, materials)	212,058
Capacity building (trainings, meetings, surveys)	74,650
Participatory evaluation (workshops, board expenses)	14,890
Communication and outreach (posters, curriculum, brochures)	15,150
Total installation cost	**421,712**

Table 2.11 *HKT hydropower operation and maintenance costs and frequency.*

Description	*Frequency*	*Cost (THB)*
Caretaker stipend	Yearly	6,362
Turbine (pump and motor)	Every 5 years	25,000
Electronics	Every 5 years	8,000

$$\text{NPC(system)} - \sum_{n=0}^{20} (\text{Sum of costs in year } n) * \left[\frac{(1+\text{inflation})}{(1+\text{discountrate})} \right]^{n}$$

Diesel generator system model

The two alternative electricity sources were modelled in addition to the hydroelectric system: a diesel generator and a connection to the local electricity grid. The diesel generator system model created for this study used a centralised 7 kVA[7] diesel generator to produce 1469 W to be available to community buildings for 85 per cent of the year. As with the HKT hydropower model, the study was not concerned with whether or not the power was consumed by the villagers, but that the power was available. It was assumed that a generator would be located in a building in the same location as the hydropower control house.

The diesel generator was assumed to have a life span of 10 years (Alsema, 2000; Fleck and Huot, 2009). The generator could produce 1 kWh of electricity from the consumption of 0.53 L of diesel fuel (Fleck and Huot, 2009).[8] The simple calculation, as shown below, resulted in the consumption of approximately 117,047 L of diesel fuel over 20 years.

[7]The size of the diesel generator chosen for the model was based on the smallest generator the author could find in Mae Sot that was rated for 24-hour production.

[8]However, if a higher fuel efficiency of 0.4 litres/kWh is used, a smaller consumption of 88, 338 lifetime litres is realised (Fleck and Huot, 2009).

Fuel consumed by generator = (total energy available to community buildings) × (diesel fuel per kWh)/(average distribution network efficiency)
= 218,977 kWh × 0.53 L/kWh/99.15 per cent
= 117,047L

The building itself would be constructed in the same fashion as the powerhouse building but would be twice the floor footprint to allow for fuel storage. The diesel generator was modelled according to the material specifications provided by Alsema (2000), which specified 30 per cent cast steel, 30 per cent steel, 35 per cent aluminium, 3 per cent plastic and 2 per cent copper. The diesel generator's weight of 275 kg was taken from industry literature and anecdotal consultation with local vendors in Mae Sot (Hoa Binh Corporation, 2010). Table 2.12 lists the important system inputs and outputs over the 20-year study time.

Items excluded from this study were generator servicing, oils and lubrications, energy used to assemble the generator from individual parts and the fuel storage tank. These exclusions would lead to slightly underestimated results for the diesel system. The same distribution grid was assumed to be used as the hydroelectric system. Diesel fuel would be transported from Bangkok to Mae Sot and then on to HKT. The system's end-of-life phase was dealt with in the same manner as for all HKT hydroelectric systems. Similarly, no materials and equipment would leave the HKT region.

In 2010, a new 7500 kVA generator costed THB 75,000. The Shell price for diesel fuel on 10 May 2010 was THB 31.98/litre (Shell Thailand, 2010). When adjusted for 3.22 per cent inflation over the previous four years, prices in 2006 THB were expected to be THB 54,638 for the generator and THB 23.3 for a litre of diesel fuel. Other diesel generator system costs included in the model were an annual maintenance stipend for the caretaker of THB 6,362 matching the stipend given in the hydropower model and a generator replacement in year 11.

Table 2.13 lists costs and replacement frequency used in a 20-year LCC for the diesel generator (Pascale *et al.*, 2011).

Table 2.12 *Diesel generator system input and outputs (including replacements).*

Item	*Description*	*Quantity*
A	Dry cement mix	400.0 kg
C	Aluminium wire with PVC sheathing	86.8 kg
O	7kVA diesel generator (per lit.: CO_2 = 1.052 kg, SO_2 = 0.235 g, CO = 0.302 g, NO_x = 1.613 g) [10]	550.0 kg
P	Diesel fuel (sulphur content, 500 ppm) [18]	117,047 litres
Q	Forest use change	23.4 m^2
R	Potential land use change from fuel leaks and spills[9]	?m^2
M	Lifetime energy available at connection to buildings	218,977 kWh

[9]The determination of land use change occurring as the result of diesel fuel spilled or leaked into the materials of the powerhouse and area surrounding it were not as easy to gauge and represent an important local ecological impact factor.

Table 2.13 *Diesel generator LCC costs and frequency.*

Description	*Frequency*	*Cost (THB)*
Installation costs	Once	54,638
Caretaker stipend	Yearly	6,362
Generator replacement	Every 10 years	54,638
Diesel fuel	Per litre	23.30

2.7 Connection to electricity grid

The grid connection scenario assumes that the power grid has reached the nearest road system turn-off point for the village and only 4 km of transmission line remains to be installed before power arrives in central HKT. The model created for this LCA consisted of 4 km of transmission line, a transformer and the distribution network in the village.

Important system inputs and outputs over the 20-year period for the connection to the grid are listed in Table 2.14. Grid electricity production for Thailand was modelled using the power grid mix from the GaBi LCA database (2006), which contained natural gas (72 per cent), brown coal (15 per cent) and large hydropower (7 per cent); which included an average Thai power grid loss of 7.3 per cent. The model did not include end-of-life considerations for materials used in the national transmission and distribution networks (GaBi 4, 2006a). The final four kilometres of transmission line was assumed to use locally sourced wooden power poles.[10] The distribution network was modelled as in the hydropower system and included a loss of 0.85 per cent.

Table 2.14 *Thai power grid connection system inputs and outputs (replacements incl.).*

Item	*Description*	*Quantity*
A	Dry cement mix	6,133 kg
C	Aluminium wire with PVC sheathing	995 kg
E	Cast Aluminium materials	235 kg
F	Ceramic materials	251 kg
H	Hand wound 380V to 230V power transformer	80.0 kg
K	Forest use change (minimum)	118 m^2
M	Lifetime energy available at connection to buildings from Thai power grid mix	218,977 kWh

[10]Transmission and distribution poles used in the Thai national grid system were usually concrete. A preliminary sensitivity analysis showed that if wooden poles in the model were replaced by concrete poles closer to the specification of Thai transmission and distribution poles, cement use would have increased by 5 per cent, thus increasing the overall system environmental impact (Pascale *et al.*, 2011).

Table 2.15 *Grid extension LCC costs and frequency.*

Description	*Frequency*	*Cost (THB)*
Grid extension	Once	2,770,046
Caretaker stipend	Yearly	6,362
Average electricity price	Per kWh	2.04

Using an inflation-adjusted (3.22 per cent) cost, grid extension cost of THB 692,512 per kilometre provided by Rapapate and Göl (2007) for rural Thai villages resulted in a four-kilometre HKT extension cost (for normal Thai grid construction using concrete poles) of THB 2,770,046. Villagers would then be required to pay the cost per kWh for electricity supplied to their village according to the Thai government–run Provincial Electricity Authority (PEA) electricity costs. Using available PEA literature, individual building metering for seven community buildings, annual price inflation of 3.22 per cent and the evenly divided consumption – per building per month – of a total 20-year energy supply of 218,977 kWh resulted in an average base per kWh electricity price of THB 2.04 (PEA, 2000). As with diesel and hydropower systems, an annual caretaker stipend of THB 6,362 was included for minor village grid maintenance. Table 2.15 lists costs and replacement frequency used in a 20-year LCC for community electrification through grid extension.

2.8 Transportation

Transportation was needed at three main points in the life cycle of the Huai Kra Thing electrification system. Transportation was needed within a country or region to transport materials required in the manufacture of the equipment and to carry the finished product to a regional international shipping hub. Transportation is then needed to carry the product from that hub to Bangkok, the central distribution point for goods in Thailand. Once in Thailand, equipment required transportation to arrive at its installed location.

Table 2.16 shows all distances and transport modes used to model the flow of items to the HKT village. The item codes used (A to P) refer to the items listed in the inventory tables above.

2.9 Results and discussion

2.9.1 Hydropower system

LCIA results per kWh for the HKT hydropower scheme are presented in Table 2.17 and in Figure 2.12.

The transmission line was the dominant component in almost all the LCIA impact categories with the exception of abiotic depletion potential (ADP) and primary energy demand (PED). The distribution network and penstock also represented a large percentage of the selected LCIA results. Powerhouse equipment contributed a larger share of PED, but saw a smaller share in all other categories. Control house equipment and the scheme's civil works both made up minor shares of all impact category results.

Table 2.16 *Transport distances and modes used for modelling of the HKT hydropower scheme (Pascale et al., 2011).*

Item	*Mode*	*From*	*To*	*Distance*	*Source*
D	Truck	Regional Japan	Tokyo	300 km	Estimate
A, 35 per cent O	Truck	RNA	Los Angeles	300 km	Estimate
B, C, E – I, 65 per cent O	Truck	RER	Rotterdam	300 km	Estimate
J	Truck	Regional Nepal	Kathmandu	300 km	Estimate
D	Ship	Tokyo	Bangkok	5,539 km	[44]
A, 35 per cent O	Ship	Los Angeles	Bangkok	14,359 km	[44]
B, C, E – I, 65 per cent O	Ship	Rotterdam	Bangkok	17,346 km	[44]
J	Plane	Kathmandu	Bangkok	2,192 km	[45]
A – J, O, P	Truck	Bangkok, Thailand	Mae Sot	495 km	[22]
A – J, O, P	Truck	Mae Sot	Huai Kra Thing	65.2 km	[22]
A – J, O, P	Foot	Huai Kra Thing	Installation Site	0–1 km	Estimate
A – J, O, P	Foot	Installation Site	End of Life	0–1 km	Estimate

Table 2.17 *Selected per kWh LCIA results for the HKT hydropower scheme model.*

Category or indicator	*HKT 3 kW hydropower scheme total*	*1. Weir, intake, canal and forebay*	*2. Penstock*	*3. Powerhouse, turbine and outflow*	*4. Transmission line*	*5. Control house and control and conditioning equipment*	*6. Distribution*
ADP (g Sb-e)	0.264	0.024	0.071	0.042	0.059	0.017	0.051
AP (g SO_2-e)	0.372	0.026	0.065	0.042	0.118	0.017	0.105
EP (g PO_2-e)	0.030	0.002	0.006	0.005	0.008	0.002	0.007
GWP (g CO_2-e)	52.7	3.7	9.8	9.0	14.7	2.7	12.9
ODP (g R11-e)	3.13E–06	4.18E–09	7.00E–09	4.18E–07	1.59E–06	7.69E–08	1.04E–06
POCP (g ethene-e)	0.030	0.002	0.005	0.003	0.010	0.001	0.009
PED (kWh)	0.150	0.015	0.046	0.029	0.025	0.011	0.023

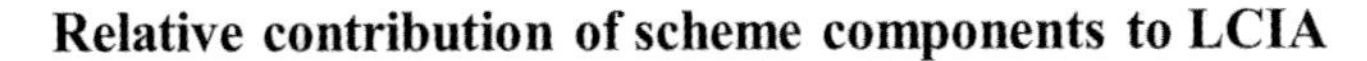

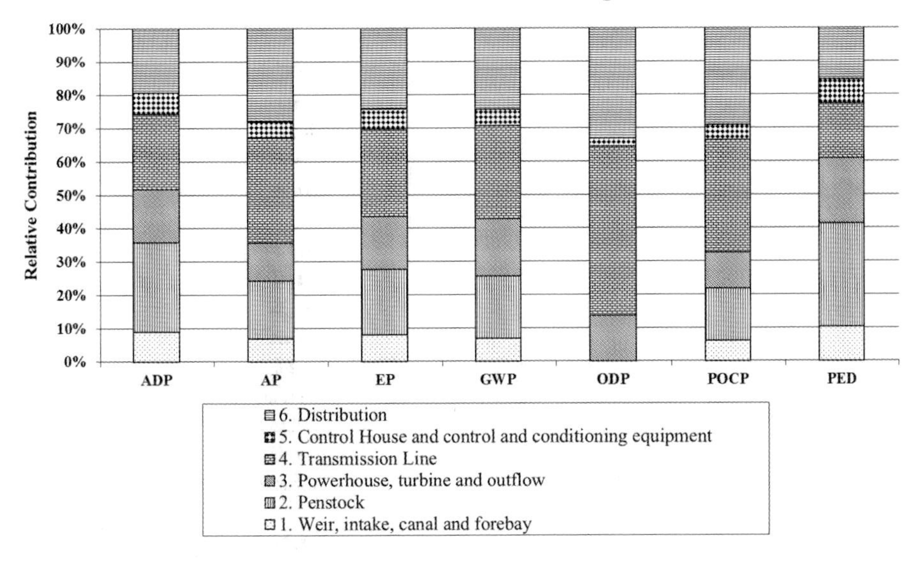

Figure 2.12 *Selected LCIA results for the HKT hydropower scheme model (Pascale et al., 2011).*

At this point, comparisons could be made with LCAs found in other hydropower and rural electrification literature. Caution should be exercised in drawing strong conclusions from the following comparisons due to variations in scope, boundary, method, impact category selection and goals between studies.

A survey of hydropower LCA literature indicated that smaller hydropower systems have higher impacts per kWh than larger systems (EPD, 2005; Dones *et al.*, 2004; Gagnon and van de Vate, 1997). With the exception of EP, the results of this study appear to support that observation.

Surveyed rural electrification LCA literature covered a centralised solar PV system, grid connection, 7.5 kVA generators running on biodiesel and diesel, two different solar home systems, a grid-connected battery charging system and a 6 kVA diesel generator (Alsema, 2000; Gmünder *et al.*, 2010). HKT hydropower system environmental impacts were consistently lower per kWh than the impacts of all rural electrification systems in the surveyed literature.

2.10 Comparison with alternative HKT village electrification options

Figure 2.13 presents a comparison of the overall per kWh LCIA results for the three HKT electrification options modelled in this LCA.

While providing the same levels of electricity and availability over a 20-year span, the modelled hydroelectric system had significantly lower values for each of the environmental impact categories considered compared to the diesel and the grid connection alternatives.

LIVERPOOL JOHN MOORES UNIVERSITY
LEARNING SERVICES

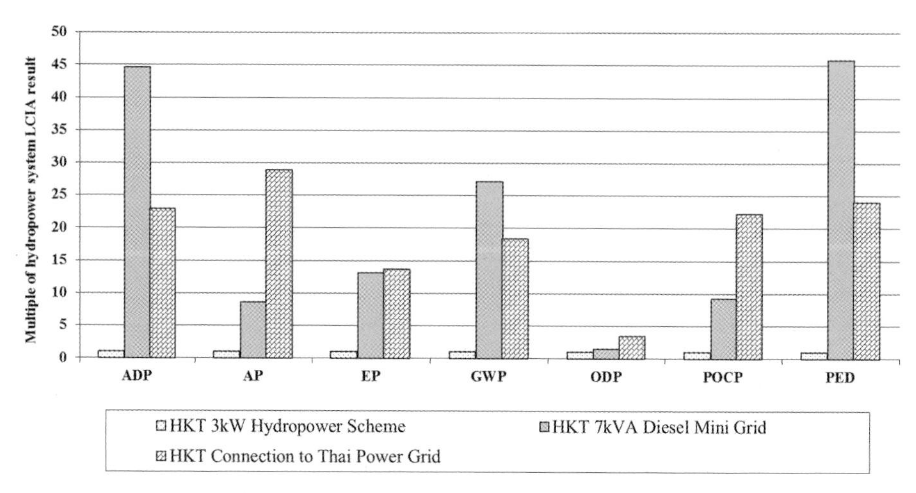

Figure 2.13 *Comparison of selected LCIA results for HKT hydropower, diesel generator and connection to the Thai power grid. The HKT hydropower system is set at one and used as a baseline (Pascale et al., 2011).*

2.10.1 Comparison with results of sensitivity analyses

A sensitivity analysis was undertaken to extend the use of these results. Two scenarios were chosen to provide ends on a wide spectrum of scenarios[11]. In the 'best case' scenario, system components lasted twice their estimated life span, the system runs 95 per cent of the year at 3 kW and no cement is used. In the 'worst case' scenario, system components last half their estimated life span, the system runs at 320 W for 85 per cent of the year and four times the cement is used. Table 2.18 presents a comparison of the best-case and worst-case scenarios with baseline hydropower, diesel system and grid connection system.

Results show that even in the worst-case scenario , the HKT hydropower system had ADP, GWP and PED environmental indicators no worse than the diesel generator or the grid connection systems. In the best case system, there were consistently returned reductions in all categories of 73 per cent or better than the baseline hydropower system.

2.10.2 Life cycle cost

The LCC results presented in Table 2.19 show that the HKT hydropower system returned the lowest total system cost of the three systems modelled in this study. Including capital costs, estimates of the in-kind donations and system operation and maintenance costs, the HKT hydropower system showed a 2006 THB total of 1,094,131. By comparison, the simplified LCC for the HKT diesel system using a fuel price adjusted for inflation but otherwise constant returned a 2006 THB total of

[11]Achieving actual system design and performance at either end was highly questionable, but served the purpose of defining a more reasonable range of design and performance for comparison.

Table 2.18 *Comparison of the LCIA results for the best and the worst case systems with alternative electrification options using multiples of baseline system's LCIA results.*

Category or indicator	*Hydropower baseline*	*7 kVA diesel generator*[12]	*Thai power grid connection*	*Hydropower best case*	*Hydropower worst case*
ADP	1	45	23	0.26	14
AP	1	9	29	0.18	20
EP	1	13	14	0.18	21
GWP	1	27	18.5	0.20	18.2
ODP	1	1.5	4	0.23	11
POCP	1	9	22	0.20	18
PED	1	46	24	0.26	14

1,645,462. The simplified LCC on the grid connection costs using an electricity price adjusted for inflation but otherwise constant returned a 2006 THB total of 3,014,742.

A sensitivity analysis involving a discount rate conducted by Pascale *et al.* (2011) showed that the hydropower system continued to return the lowest LCC until the discount rate was above 17 per cent. Furthermore, 'although inclusion of unaccounted for costs in the NPC calculation was expected to increase each system's total cost, missing costs in the hydropower system were few and expected to have the least impact. The costs missing from the NPC calculations of the diesel mini grid system were many and expected to have the most impact on its NPC, thus increasing the hydropower system's clear 20 year financial advantage over its closest competitor for this location.'

Table 2.19 *NPC 20-year cost inputs and calculation results for each base system.*

System costs over 20 years	*Hydropower base system*	*7 kVA diesel generator*	*Thai power grid connection*
Installation, purchase or extension	THB 973,587	THB 54,638	THB 2,770,046
Annual caretaker stipend	THB 6,362	THB 6,362	THB 6,362
Equipment replacement costs – years 6 and 16	THB 33,000	n/a	n/a
Equipment replacement costs – year 11	THB 33,000	THB 54,638	n/a
Diesel fuel cost per litre, 2006 (adjusted from Shell Thailand, 2010)	n/a	THB 23.30	n/a
Electric utility cost per kWh[13]	n/a	n/a	THB 2.04
Calculated 20-year net present cost	**THB 1,094,131**	**THB 1,645,462**	**THB 3,014,742**

[12]If 0.4 litre/kWh consumption figures are used, ADP = 34, AP = 8, EP = 12, GWP = 26, ODP = 1.2, POCP = 8, PED = 35.

[13]Calculated using Ref 43 and based on evenly divided consumption and individual building metering.

2.10.3 Design considerations and limitations on the use of the results

The modelling and comparison of rural electrification schemes using approximate and substitute processes and estimated costs limits the use of LCA and LCC results to a specific time and list of assumptions. Only continual LCA and LCC iterations involving the integration of Thailand and product-specific processes and updated costs as they become available would allow tools such as LCA and LCC to retain their relevance for future designers and decision makers.

The exclusion of changes in land use from the formal LCIA results seemed to be a common theme in recent industry hydropower LCAs such as Vattenfall (2008) and in recent rural electrification LCA literature such as Gmünder *et al.* (2010). As land use directly impacts on the local environments and the health of local environments directly impacts upon local people, it is an important environmental aspect of rural electrification and should be a central part of LCA reporting. Uncontaminated land also holds a major potential financial value for local people and land use change should be included formally in LCC reporting and results.

The benefits arising from any design-improving action depend on the manner in which change was implemented and the system-wide effects of changing a single aspect of the system. For example, the sensitivity analysis of the HKT system indicated that removing cement from the system could reduce GWP potential by 25 per cent. However, if the reduction in cement could also make the system less robust and decreased the system performance significantly, the net system GWP per kWh could actually rise. Conversely, if doubling in the amount of cement used in the scheme pushed the system performance towards an optimum operation, the net GWP had the potential to decrease per kWh with respect to the baseline system.

LCA and LCC modelling and presentation of results occurred in a very technical framework designed for the characteristics and operation of the hydropower system. No effort was made in this study to assess fully and incorporate the qualitative and social benefits and shortcomings of the hydropower or alternative electrification systems. Recognition of the problem of quantitatively comparing electrification systems with qualitative differences was noted by both Rule *et al.* (2009) and Gagnon *et al.* (2002).

2.11 Conclusion

The use of LCA on the HKT hydropower system demonstrated that the system could yield better environmental outcomes than the diesel and grid extensions alternatives. The study resulted in the enumeration of the environmental credentials of the HKT hydropower system. Credentials were reinforced through comparison with rural electrification alternatives, assessment of results against existing LCA literature and sensitivity analyses that allowed an insight into the basis for credentials.

The use of a simple LCC on the HKT hydropower system demonstrated the system's clear financial advantage over the 20-year LCA span. According to the results of the financial analysis, the HKT hydropower system returned the lowest total system cost when compared with the modelled alternatives.

The study also suggests that stakeholder involvement is important during all LCA phases in order to adapt the LCA to local conditions, normalise the results for country-specific use and report findings in an appropriate manner for local populations. Undertaken collaboratively with local stakeholders and integrated with additional tools such as a detailed social impact assessment, LCA and LCC can work powerfully together to aid rural electrification designers and decision makers.

References

Allderdice, A. and J.H. Rogers (2000) *Renewable energy for microenterprise*. National Renewable Energy Laboratory, Golden, CO. Available from: www.nrel.gov/docs/fy01osti/26188.pdf.

Alsema, E.A. (2000) *Environmental life cycle assessment of solar home systems. Report NWS-E-2000-15*. 2000, Department of Science, Technology and Society, Utrecht University: Utrect, The Netherlands. Available from: http://www.chem.uu.nl/nws/www/publica/Publicaties2000/e2000-15.pdf.

AutoNavi and Google (2014) *Google maps engine* [cited 2014 18 August]. Available from: https://mapsengine.google.com/map/edit?mid=zPzakWsfS1VM.k5k6whpV_zMU.

BGET (2006) *Huai Kra thing micro-hydro survey 08 Nov. 2005* [cited 2010 23 January].

Blanco, C.J.C., Y. Secretan and A.L.A. Mesquita (2008) Decision support system for micro-hydro power plants in the Amazon region under a sustainable development perspective. *Energy for Sustainable Development*, **12**(3), 25–33.

Bureau of Trade and Economic Indices (2010) *General consumer price index of country (Y/Y)*. [cited 2010 19 March]. Available from: http://www.price.moc.go.th/price/cpi/index_new_e.asp.

Distances.com (2010) *World ports distances calculator*. n.d. [cited 2010 1 March]. Available from: http://www.distances.com/.

Dones, R., T. Heck, and S. Hirschberg (2004) *Greenhouse gas emissions from energy systems: comparison and overview*, in *PSI annual report annex IV*. Paul Scherrer Institute, Villigen. 27–40.

Fleck, B. and M. Huot (2009) Comparative life-cycle assessment of a small wind turbine for residential off-grid use. *Renewable Energy*, **34**(12), 2688–2696.

GaBi 4 (2006) *Dataset documentation to the software-system and databases*. [cited 2010 27 April]. Available from: http://gabi-software.com/supp.html.

Gagnon, L., C. Bélanger, and Y. Uchiyama (2002) Life-cycle assessment of electricity generation options: The status of research in year 2001. *Energy Policy*, **30**(14), 1267–1278.

Gagnon, L. and J.F. van de Vate (1997) Greenhouse gas emissions from hydropower: the state of research in 1996. *Energy Policy*, **25**(1): 7–13.

Gerbens-Leenes, P.W., A.Y. Hoekstra, and T.H. Van der Meer (2008) *Water footprint of bio-energy and other primary energy carriers*, in *Value of water research report series No. 29*, UNESCO-IHE Institute for Water Education, Delft, the Netherlands. Available from: http://www.waterfootprint.org/Reports/Report29-WaterFootprintBioenergy.pdf.

Gmünder, S.M., *et al.* (2010) Life cycle assessment of village electrification based on straight jatropha oil in Chhattisgarh, India. *Biomass and Bioenergy*, **34**(3), 347–355.

Greacen, C. (2004) The marginalization of “small is beautiful”: micro-hydroelectricity, common property, and the politics of rural electricity provision in Thailand, in *Energy and Resources Group*. 2004, University of California Berkley. Available from: www.palangthai.org/docs/GreacenDissertation.pdf.

Greacen, C. (2006) *Project report – Huai Kra thing micro-hydro project*. Border Green Energy Team, Mae Sot.

Greacen, C., and M. Kerins (n.d.) *A guide to pump-as-turbine pico-hydropower systems*: Palang Thai.

Guinee (ed), J.B. *et al.* (2001) *Life cycle assessment: an operational guide to the ISO standards*. J.B. Guinee, editor. Ministry of Housing, Spatial Planning and the Environment and Centre of Environmental Science – Leiden University. Available from: http://cml.leiden.edu/research/industrialecology/researchprojects/finished/new-dutch-lca-guide.html.

Hoa Binh Corporation (2010) *DG 7500*. n.d. [cited 2010 9 May]. Available from: http://hbc.com.vn/index.php?page=shop.product_details&flypage=flypage_lite_pdf.tpl&product_id=244&category_id=61&option=com_virtuemart&Itemid=70&lang=en.

IEA (2002) Environmental and health impacts of electricity generation: a comparison of the environmental impacts of hydropower with those of other generation technologies, in *Implementing agreement for hydropower technologies and programmes*. Available from: www.ieahydro.org/reports/ST3-020613b.pdf.

International Standards Organization (1997) *Environmental management – life cycle assessment – principles and framework (ISO 14040: 1997)*. Geneva.

International Standards Organization (2010) *ISO 14044:2006*. 2010 [cited 2010 9 April]. Available from: http://www.iso.org/iso/catalogue_detail.htm?csnumber=38498.

Inversin, A. R. (1986) *Micro-hydropower sourcebook*. NRECA International Foundation, Arlington, VA.

Kammen, D.M. and C. Kirbui (2008) *Poverty, energy and resource use in developing countries: focus on Africa*. New York Academy of Science, 348–357.

Kenfack, J., *et al.* (2009) Microhydro-PV-hybrid system: sizing a small hydro-PV-hybrid system for rural electrification in developing countries. *Renewable Energy*, **34**(10), 2259–2263.

Khennas, S., *et al.* (2000) *Best practices for sustainable development of micro hydro power in developing countries: final synthesis report contract R7215*. 2000, Department for International Development and The World Bank. Available from: www.microhydropower.net/download/bestpractsynthe.pdf.

KNCE (2007) *Final report (างนดคงา): THA-05-08 Clean energy*.

Lynch, A., *et al.* (2006) *Threatened sustainability: the uncertain future of Thailand's solar home systems*. BGET, Mae Sot.

Maher, P., and N. Smith (2001) *Pico hydro for village power: a practical manual for schemes up to 5 kW in hilly areas*. Available from: http://www.eee.nottingham.ac.uk/picohydro/docs/impman(ch1-6).pdf (accessed 29 April 2010).

Moore, A.D. (2009) *Life cycle assessment (LCA) of a 1kWp photovoltaic system installed in Australia*, in *School of engineering and energy*. Murdoch University, Murdoch, Australia.

OECD (2009) *Environment: water consumption*, in *OECD Factbook 2009*.

Pascale, A. (2010) *Life cycle analysis of a community hydroelectric system in rural Thailand*, in *School of engineering and energy*. Murdoch University, Murdoch, Western Australia. Available from: http://researchrepository.murdoch.edu.au/3486/.

Pascale, A., T. Urmee, and A. Moore (2011) Life cycle assessment of a community hydroelectric power system in rural Thailand. *Renewable Energy*, **36**: 2799–2808.

PEA (2000) *Electricity rates*. [cited 2010 10 May]. Available from: http://www.pea.co.th/th/eng/downloadable/electricityrates.pdf.

Pereira, M.G., M.A.V. Freitas, and N.F. da Silva (2010) Rural electrification and energy poverty: empirical evidences from Brazil. *Renewable and Sustainable Energy Reviews*, **14**(4), 1229–1240.

Rapapate, N., and Ö. Göl (2007) Use of photovoltaic systems for rural electrification in Thailand. *International Conference on Renewable Energies and Power Quality*, Sevilla, Spain.

Rule, B.M., Z.J. Worth, and C.A. Boyle (2009) Comparison of life cycle carbon dioxide emissions and embodied energy in four renewable electricity generation technologies in New Zealand. *Environmental Science & Technology* **43**(16), 6406–6413.

Sein, S.L. (2006) *Karen village Mae Sot:* KNCE.

Shell Thailand (2010) *Fuel price*. [cited 2010 10 May]. Available from: http://www.shell.co.th/home/page/tha-en/products_services/on_the_road/fuels/fuel_price/app_fuel_prices.html.

TISI (2010) *Thai Industrial Standards Institute [TISI]*. n.d. [cited 20 May 2010]. Available from: http://www.tisi.go.th/eng/index.php?option=com_content&view=article&id=2&Itemid=4.

UNDP (2003) *Human development report 2003: millennium development goals: a compact among nations to end human poverty*. New York. Available from: http://hdr.undp.org/en/reports/global/hdr2003/chapters/.

Urmee, T.P. and D. Harries (2009) A survey of solar PV program implementers in Asia and the Pacific regions. *Energy for Sustainable Development*, **13**(1), 24–32.

Urmee, T.P., D. Harries and A. Schlaprfer (2009) Issues related to rural electrification using renewable energy in developing countries of Asia and Pacific. *Renewable Energy*, **34**(2),354–357.

Varun, I.K.B. and R. Prakash (2009) LCA of renewable energy for electricity generation systems – a review. *Renewable and Sustainable Energy Reviews*, **13**(5), 1067–1073.

Vattenfall (2008) *Vattenfall AB generation Nordic certified environmental product declaration EDP of electricity from Vattenfall's Nordic hydropower* 2008. Available from: http://www.environdec.com/en/Detail/?Epd=7468#.U_KYh6P_Rvs.

WebFlyer (2010) *Mileage calculator*. n.d. [cited 2010 1 March]. Available from: http://www.webflyer.com/travel/mileage_calculator/.

Williams, A. (1995) *Pumps as turbines: a users guide*. 1997 ed. Intermediate Technology Publications, London.

World Bank, E., and International Energy Agency (2013) *Global tracking framework – sustainable energy for all*, in *Working paper 77889*, S.B. Kennedy, editor. World Bank, Washington, DC. Available from: http://www.iea.org/publications/freepublications/publication/Global_Tracking_Framework.pdf.

XE.COM (2013) *Current and historical rate tables* [cited 2013 5 December]. Available from: http://www.xe.com/currencytables/?from=THB&date=2006-02-01.

Chapter 3
Selection indicators for stabilisation of pavement systems

Filippo G. Praticò, Sireesh Saride and Anand J. Puppala

Summary: In this chapter, the application of life cycle cost analysis (LCCA) on optimising and selecting the most appropriate stabiliser and stabilisation method is discussed through a new conceptual engineering-economic model. In this new LCCA model analysis, agency, user and other costs are addressed. The model is demonstrated and validated through two case studies on low-volume roads (LVRs): one from Italy and the other from the United States. Results from both the case studies demonstrated that under specific boundary conditions, soil stabilisation can play an important role, merging the environmental and the mechanical effectiveness of LVRs.

3.1 Introduction and background

It is always a crucial task for any transportation agency to select the most appropriate stabilisation technique for a given field scenario. To this end, a promising tool known as life cycle cost analysis (LCCA) has been widely used in the decision-making process. LCCA is an engineering-economic analysis tool which compares the relative merits of competing project implementation alternatives. Minimising the pavement life cycle costs (present worth value, PWV or PV, and the equivalent uniform annual cost) will increase the sustainability of the pavement system to provide a prolonged service life and to reduce user impact costs (Douglas and Molenaar, 2004).

Much work has been dedicated to the development of LCCA-based procedures and models (Walls and Smith, 1998; Wilde *et al.*, 1999; FHWA, 2002; Ou and Swarthout, 1986; Abourizk and Ariaratnam, 2003; Gerbrandt and Berthelot, 2007), but limited data are available on LCCA models for pavements where soil stabilisation techniques are involved. Hence, in this chapter, a new simple conceptual engineering-economic tool, based on LCCA, is developed to optimise the best alternative infrastructure investment option for pavements. The new model has been validated with data from two low-volume roads (LVRs), one each from Italy and the United States.

3.2 Model and methodology

An engineering-economic model is developed based on three prevailing options at a specific pavement construction site for selecting a stabiliser/stabilisation technique. The three following scenarios are considered for analysis:

(a) Maintaining *in situ* soils as they are (i.e. no soil stabilisation, NS, the 'zero' option);
(b) Stabilising the *in situ* soil (S);
(c) Substituting the *in situ* soil (excavation and replacement soil substitution, SUB).

The notations NS, S and SUB describe the corresponding scenarios. For each case, three main classes of expenditure are analysed:

1. Agency Costs (AC) includes initial construction costs, successive rehabilitations and maintenance costs;
2. User Costs (UC) relates to traffic delays, vehicle operating costs, accident and safety costs;
3. Environmental Externality Costs (EX) relates to pollution, environmental depletion, global climate change, noise cost and others.

It should be noted that the AC are affected by the traffic levels due to dependence of expected life on the number of axle loads. Further, delays and vehicle operating costs depend on the traffic loads. The externality costs (Yin and Siriphong, 2006) are difficult to quantify whether or not traffic is considered as a main parameter.

3.2.1 Model development

3.2.1.1 Expected life

A conceptual framework of the proposed model is presented in Figure 1, with the nomenclature being given in Table 1. The model set out in this study is based on simple equations (3.1) to (3.20). Equation (3.1) refers to the expected life of the pavement structure. When the American Association of State Highway and Transportation Officials (AASHTO, 1993) approach is used, wheel loads of various magnitudes and repetitions ('mixed traffic') are converted to an equivalent number of 'standard' single axle loads (ESALs) of 80 kN. The number of ESALs that a specific pavement will experience over its design life is then compared to the ESALs the pavement can withstand. The new mechanistic-empirical pavement design guide, MEPDG (2007), adopts a load spectra approach and therefore an expected life can be derived for the 'zero' hypothesis (i.e. control or no stabilisation) and for the alternative (stabilisation). Further, the pavement geometry (G_e) and mechanical properties (M), for a given traffic and climatic conditions, will give a different expected life (E) of the pavement.

$$E = f(M, G_e) \tag{3.1}$$

3.2.1.2 Agency costs

The same pavement management approach will lead to different financial and construction consequences for the two options, stabilisation and substitution, considered. The initial construction cost (C) will be different, as well as the construction period (n_k) and the costs incurred in successive rehabilitations (C_{REk}) are different. Initial construction costs, for the soil stabilisation case, will be higher than 'zero' option. This will affect the present value of the agency costs (PV_{ACT}) estimated for a period of analysis 'T' years as shown in equation 3.2.

$$PV_{ACT} = C + \sum_{k=1}^{N} \frac{C_{REk}}{(1+i)^{n_k}} - \frac{S_a}{(1+i)^{n_N}} \tag{3.2}$$

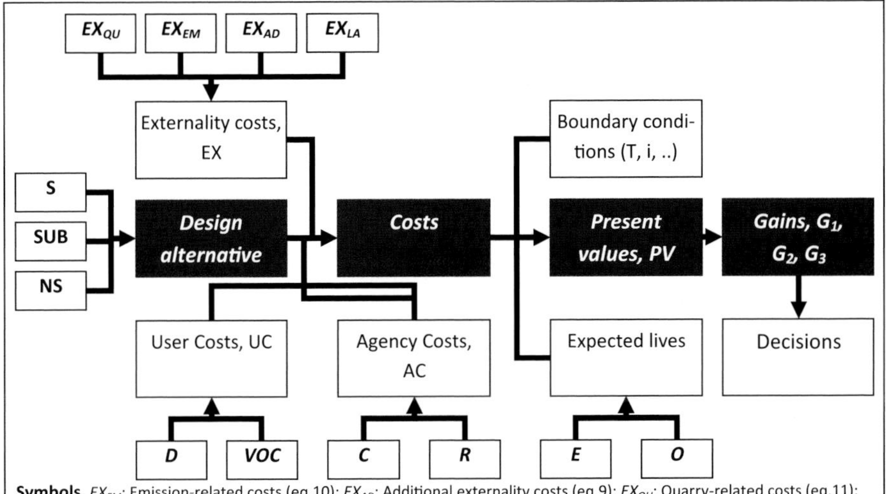

Symbols. EX_{EM}: Emission-related costs (eq.10); EX_{AD}: Additional externality costs (eq.9); EX_{QU}: Quarry-related costs (eq.11); EX_{LA}: Landfill-related costs (eq.12); *VOC*: Vehicle operating costs (eq.5); *D*: Delay-related costs (eq. 6); *R*: Rehabilitation costs (eq.4); *C*: Construction Costs (eq.4); *E*: expected life; *O*: expected life of successive rehabilitations; S: soil stabilization; SUB: soil substitution; NS: neither S nor SUB; if $G_1=PV_{NS}-PV_S>0$ then S is better than NS; if $G_2=PV_{SUB}-PV_S>0$ then S is better than SUB; if $G_3=PV_{SUB}-PV_{NS}>0$ then NS is better than SUB.

Figure 3.1 *Schematic of the proposed LCCA model (Praticò et al., 2011b).*

Table 3.1 *Factorial plan of experiments and LCC analysis options.*

Pavement type	Rigid pavement (JPCP, jointed plain concrete pavement)
Design alternatives	Control (NS), soil stabilization (S), soil substitution (SUB)
Main variables	Cost of soil stabilization, AADT
Main indicators	PV_{AC}, PV_{UC}, P_{VEX}, G_{AC1}, G_{UC1}, G_{EX1}, G_{AC2}, G_{UC2}, G_{EX2}
Miles	4
Analysis period	Infinite
Discount rate	4%
Beginning of analysis period	2009

where i is the interest rate, n_k is the year in which the kth rehabilitation will take place, $n_N \leq T$ (years) and S_a (salvage value of the pavement) can be estimated as a fraction of C_{REk}.

To provide flexibility to extend this model to a more complex analysis or to a high-volume road, higher periods of analysis (T) are required. In such cases, the period of analysis (T) in which the pavement life cycle cost is analysed can be considered as infinite (Praticò, 2007). In this case, it will follow

$$\lim_{T\to\infty} \frac{S_a}{(1+i)^{n_N}} = 0 \qquad (3.3)$$

If n_1=E, $n_2 = E + O,\ldots, n_k$=$E + (k - 1)O$ (where E and O are expected lives, years), and road surface properties are complying (Praticò, 2007), then the following expression can be derived:

$$PV_{AC} = C + \frac{C_{RE}}{(1+i)^E} \cdot \left(1 - \frac{1}{(1+i)^O}\right)^{-1} \tag{3.4}$$

where PV_{AC} is the present value (PV) of the agency costs (AC) when $T \rightarrow \infty$. The same conceptual framework can be applied to estimate routine maintenance costs.

3.2.1.3 User costs

User costs (UC) are related to delays (D), originated by work zones (WZ) of given length and duration and vehicle operating costs (VOC). Their effects depend not only on the geometric features, but also on the interaction between traffic demand and the number of lanes of given width, etc. The work zone duration (WZD_k) of successive maintenance and rehabilitation operations for the zero option (NS) will usually be shorter than the soil stabilisation case (S), if the stabilisation is considered as a part of rehabilitation operations. This fact will affect the UCs and their present value (PV_{UC}, see equations 3.5 and 3.6). When the VOC are taken into account, due to the increased mechanical properties of the stabilised subgrade (S), the roughness in the zero option (NS) (control) will probably be higher and therefore the VOC will be considerably higher (equations 3.5 and 3.6). It is very probable that in the case of S, the time between successive rehabilitations will increase, while each rehabilitation operation will affect the WZD (equation 3.6), added time (ΔT) and stopping costs (SC).

For practical derivation of present value related to delays (PV_D) and vehicle operating costs (PV_{VOC}), it is relevant to observe that the time–cost relation is an important subject and affects the social impact of public budget management. It results in the following:

$$PV_{UC} = PV_D + PV_{VOC} \tag{3.5}$$

$$PV_D = f(WZL_k, WZC_k, WZD_k, WZSL_k NLO_k, TD_k, VUC_k, TOD_k, \Delta T_k, SC_k) \tag{3.6}$$

where PV_{UC} stands for the present value of user costs (\$); PV_D stands for the present value related to delays (\$); PV_{VOC} stands for the present value of vehicle operating costs (\$); WZL_k stands for the length of lanes in the work zone area for the kth rehabilitation process; WZC_k stands for the work zone capacity; WZD_k refers to the duration; NLO_k is the number of lanes open; $WZSL_k$ is the work zone speed limit; TD_k stands for the traffic data; VUC_k is the value of user costs; TOD_k is the traffic hourly distribution; ΔT_k is the added time and SC_k refers to the stopping costs. In the case of LVRs, the estimation of PV_D is a complex exercise. Usually traffic is intermittently reduced to a single lane, alternating directions using flaggers. The corresponding operating speeds are crucial in the derivation of PV_D. However, the user costs, being related to the delays and the operating costs, can have a negligible effect when compared to the agency costs, mainly due to the low traffic volume on LVRs.

3.2.1.4 Environmental cost

The most crucial impacts of the transportation system on the environment are related to climate change, air quality, noise, water quality, soil quality, biodiversity, land take, quarries, landfills and visual impacts (Yin and Siriphong, 2006; Willis and Garrod, 1999; Olof, 1997; Ian *et al.*, 2009). The sustainable design of pavements can considerably reduce the environmental costs. Several researchers have shown the successful reuse of recycled materials in pavement construction (Nunes *et al.*, 1996; Chesner *et al.*, 2001; Ellis, 2003; Saride *et al.*, 2010; Praticò, 2004). The quantification of environmental costs is difficult, however, if some simplifications and generalisations are assumed, this objective can be pursued. In addition, extensive laboratory and field experiments will answer such questions related to environmental influences on the total project budget. Based on this discussion, the following four main classes of environmental costs, EX, are considered, both in the construction (EX_0) and in the successive kth rehabilitations (EX_k): emission-related (EX_{EM}), quarrying-related (EX_{QU}), landfill-related (EX_{LA}) and additional (EX_{AD}) costs. The following equation is derived for the present value of the environmental costs for a period of analysis equal to T:

$$PV_{EXT} = EX_0 + \sum_{k=1}^{N} \frac{EX_k}{(1+i)^{n_k}} \tag{3.7}$$

When T tends to infinite, the following equation can be derived:

$$PV_{EX} = EX_0 + \frac{EX_k}{(1+i)^E} \cdot \left[\frac{1}{1 - \frac{1}{(1+i)^O}} \right] \tag{3.8}$$

$$EX_k = EX_{kEM} + EX_{kQU} + EX_{kLA} + EX_{kAD} \quad \text{for } k = 0, 1, \ldots N. \tag{3.9}$$

If congestion, road accidents, local air pollution and global warming are taken into account, based on current literature (Olof, 1997; Ian *et al.*, 2009; Praticò *et al.*, 2010), fuel consumption, FC (l/km) in this scenario is also a key factor. The following tentative expression can be derived for this condition:

$$EX_{kEM} = \sum_j \frac{VN_{kj}}{CA_{kj}} \cdot D_{kj} \cdot EM_{kj} \cdot FC_{kj} \tag{3.10}$$

where the subscript 'k' refers to the kth rehabilitation process and the subscript 'j' refers to the given process material, VN_{kj} (m³) is the volume of material needed (e.g. slab/soil/stabiliser), CA_{ki} is the average truck capacity (m³), D_{kj} is the average distance to cover (round trip, km), EM_{kj} is an emission surcharge factor ($/l) and FC_{kj} stands for fuel consumption (l/km).

For quarrying/production (for a given j-type material/process) environmental cost, the following expression is proposed:

$$EX_{kQU} = \sum_j VN_{kj} \cdot QU_{kj} \tag{3.11}$$

where VN_{kj} (m³) is the volume of material needed (slab/soil/stabiliser) and QU_{kj} takes into account the environmental cost quantification (\$/m³).

Finally, landfill-related environmental costs are considered through the following expression:

$$EX_{kLA} = \sum_j VN_{kj} \cdot LA_{kj} \tag{3.12}$$

where VN_{kj} (m³) is the volume of material needed (slab/soil/stabiliser) and LA_{kj} takes into account the environmental cost quantification (\$/m³).

3.2.1.5 Present value and gain

The final objective of this model is to evaluate each alternative with respect to another in terms of total costs and net benefits. To quantify the benefits in terms of present value, an indicator '*gain*' is introduced. This section sums up the model described above.

AC, UC and EX contribute to the overall present value, PV given as

$$PV = PV_{AC} + PV_{UC} + PV_{EX} \tag{3.13}$$

Based on the three different costs (AC, UC, EX), it is possible to estimate the gain (G) originated by soil stabilisation (S) with respect to the control case (NS) termed as gain G_1, or with respect to the case in which *in situ* soil is removed and substituted (SUB) termed as gain G_2. The gain can be evaluated with the consideration of only agency costs (e.g. G_{AC1}), only environmental costs (e.g. G_{EX1}) or any other combinations as appropriate (see Figure 3.1). By referring to the comparison of no stabilisation (NS) against stabilisation (S), the gain will be evaluated as

$$\begin{aligned} G_1 &= \left(PV_{AC} + PV_{UC} + PV_{EX}\right)_{NS} - \left(PV_{AC} + PV_{UC} + PV_{EX}\right)_S \\ &= \left[\left(PV_{AC}\right)_{NS} - \left(PV_{AC}\right)_S\right] + \left[\left(PV_{UC}\right)_{NS} - \left(PV_{UC}\right)_S\right] + \left[\left(PV_{EX}\right)_{NS} - \left(PV_{EX}\right)_S\right] \\ &= G_{AC1} + G_{UC1} + G_{EX1}. \end{aligned} \tag{3.14}$$

Further, it is easy to demonstrate that the gain in choosing the NS solution with respect to the solution SUB, hereafter termed as G_3, is given by

$$G_3 = G_2 - G_1. \tag{3.15}$$

Even if the concept of perpetual pavements is of growing interest (Timm *et al.*, 2006), for the design of LVRs the following further boundary conditions are usually used:

$$C \le C_{max} \tag{3.16}$$

$$C_{REk} \le C_{REkmax} \tag{3.17}$$

where C_{max} (maximum construction cost) and C_{REkmax} (maximum rehabilitation cost) refer to the budget potential of the local road authorities.

Then, equation 3.16 usually yields the following:

$$E, O \leq 15\text{–}45 \text{ years,} \quad (3.18)$$

Further, with reference to traditional analyses, the following equation can be assumed as a key factor in pavement management terms:

$$G_{AC2} > 0. \quad (3.19)$$

Also, with reference to short-term analysis, the cost of stabilisation alternatives would usually be higher than for the control case. It follows:

$$C_S > C_{NS} \quad (3.20)$$

Following the model set out above, the methodology can be summarised in the following phases:

i. *First phase:* formulating alternatives, boundary conditions and estimating the expected lives.
ii. *Second phase:* project-level inputs (analysis options, traffic data, value of user time, traffic hourly distribution, added time and vehicle stopping costs).
iii. *Third phase:* alternative-level inputs (agency construction costs, work zone lengths for each alternative).
iv. *Fourth phase:* deriving present values and conveniences.

3.2.2 Advantages and limitations of the model

The current LCCA model is developed based on a simple conceptual framework. This model can consider any maximum life of analysis period. This option will provide more flexibility to the model if it is required for cost–benefit analysis of a high-volume road such as interstate highway for a longer analysis period.

However, adequate care must be exercised while deriving the environmental costs for a given scenario. Reasonable assumptions are needed if the relevant laboratory and field data are unavailable. Further developments and improvements are necessary to incorporate synergistic environmental cost calculations and their implementation in the model.

3.3 Validation of the model

3.3.1 Introduction to low-volume roads

LVRs can be broadly defined as roads with low traffic volume such as unpaved roads, where the annual average daily traffic (AADT) is less than 500 vehicles per day, and paved roads, where the AADT is higher than 500 vehicles per day (Muench *et al.*, 2004; Behrens, 1999; Keller and Sherar, 2003). In the United States, farm to market (FM) roads come within this category. These roads generally connect low-traffic-density towns and villages with major cities and are very much responsible for the socioeconomic development of rural communities (Maryvonne, 2007; Behrens, 1999; Muench *et al.*, 2004). According to AASHTO (1993), LVRs can be defined as paved roads with approximately one million ESALs in a given performance period, with

50,000 ESALs as the practical minimum. Washington State department of transportation (WSDOT) limits to 50,000 ESALs/year for the design of LVR for a design period of 20 years (WSDOT, 1995).

The LVRs are designed for shorter durations based on the traffic volume. However, if the subgrade condition is weak, the design life of an LVR is considerably shortened and requires frequent maintenance and rehabilitation. It was estimated that in the United States the annual maintenance and rehabilitation costs, exclusively for LVRs, was $82 million (Behrens, 1999). Methods for increasing the service life of LVRs include treatment with calcium-based stabilisers, non-calcium-based stabilisers, asphalt stabilisation and geosynthetic reinforcement (Munech, *et al.*, 2004; Bushman *et al.*, 2005; Kota, 1996; Sherwood, 1993; Bugge and Bartelsmeyer, 1961; Croft, 1967; Little, 1995; Puppala *et al.*, 1998; Basma and Tuncer, 1991). Puppala and his group have ranked the best stabilisers for use on LVRs constructed in sulphate-rich expansive soils (Puppala *et al.*, 2003, 2007; Sirivitmaitrie, 2008).

To evaluate the new LCCA model, two case studies, one from Italy and the other from Texas in the United States, are analysed and presented in this section. The case studies were chosen such that both had some commonalities, for example a similar pavement structure. Nomenclature, the factorial plan of experiments and the general analysis options considered for both cases are presented in Tables 3.1 and 3.2. For the purpose of analysis, three stabilisation costs and three traffic densities are considered within the range of actual values.

3.3.2 Case Study I – Southern Italy

In this case, a two-layer rigid pavement system (slab on lime-stabilised subgrade) was considered. The information required for the analysis was collected from laboratory studies and from the construction of field test sections. The laboratory tests included classification tests, with California bearing ratio (CBR) and resilient modulus (Mr) tests conducted on the subgrade soils both before and after the stabilisation process. The subgrade soils were found to be soft clays and classified as CL according to the unified soil classification system (USCS). The test results are presented in Table 3.2. It can be seen that the stabilisation process improved the mechanical behaviour of the stabilised soils. For analysis purposes, three different stabilisation costs (Stb.cost (€/m^3): 4, 8 and 16, as shown in Figure 3.2) and three traffic densities (500, 1000 and 2000 vehicles/day, as shown in Figure 3.2) were considered within the range of actual values. For environmental cost calculations, the quarrying (QU) and land fill (LA) surcharges were considered as 1.95 €/ton. The fuel consumption surcharge factor was considered as 0.7 €/l. The actual values are presented in Table 3.2.

3.3.2.1 Expected life

To estimate the expected life of the pavement, the MEPDG and the AASHTO Design Guide (AASHTO, 1993; MEPDG, 2007; Kim and Newcomb, 1991; Hall and Bettis, 2000; Praticò *et al.*, 2011a, 2011b) methods are used. A 75-per cent design reliability level was assumed for the two-layered rigid pavement system (Table 3.2). The environmental conditions corresponding to southern Italy climate were considered. Three different design alternatives – (a) no stabilisation/control

Table 3.2 *Summary of parameters (untreated vs. treated), for Southern Italy, and for Arlington, Texas (United States).*

Location	*Southern Italy*	
Treatment condition (in this case a lime treatment was used)	*Untreated*	*Treated*
Resilient modulus (M_r), MPa	69	179
k, pci MN/m³ [1pci = 0.283 N/cm³]	45.4	95
PCC E_c, MPa [1psi = 0.69N/cm²]	27,789.38	27,789.38
Project description: Application areas (m²) [1SY = 0.836m²]	64,000	64,000
Project description: Surface (mm)/ stabilized subgrade (mm)	152/200	152/200
Project description: Width of street (m)/length of Street (m)	10/6,400	10/6,400
Routine maintenance: Seal @ 5 yrs, 10 yrs, etc. (€/m²)	2.6	2.6
Routine maintenance: Resurfacing 2" @10 yrs (€/m²)/ resurfacing 2" @20yrs (€/m²)	8.5/8.5	–/8.5
Construction cost: Jointed Plain Concrete Pavement (JPCP)(6") (€)	1,920,000	1,920,000
Construction cost: Stabilization (material), (€) + stabilization (labor), (€)	0	227,585
Traffic/ work zone data: AADT construction year (total for both directions)	500–2,000	
Traffic/ work zone data: Single unit trucks/buses as percentage of AADT	10%	
Traffic/ work zone data: Annual growth rate of traffic	4%	
Traffic/ work zone data: Lanes open in each direction under normal/ work zone conditions	1/ Two way lane shift (½)	
Traffic/ work zone data: Work zone duration (days)	20	27
Value of user time: Average value of time for passenger cars/trucks/buses	15 (€/hour)	
Location	*Arlington, Texas, United States*	
Treatment condition (in this case a lime-cement treatment was used)	*Untreated*	*Treated*
USCS classification/ free vertical swell, %/ linear shrinkage, %	CH/6.3/20.2	CL/0.0/0.0
UCS, kPa/ resilient modulus (M_r), MPa/ k, MN/m³	87.0/42/29	1,730.0/244/126.5

(Continued)

Table 3.2 *Continued*

Location	*Arlington, Texas, United States*	
Treatment condition (in this case a lime-cement treatment was used)	*Untreated*	*Treated*
PCC E_c, MPa	27,789.38	4,030,509
Project description: Application areas (m²)	64,000	64,000
Project description: Surface (mm)/stabilized subgrade (mm)	200/200	200/200
Project description: width of street (m)/length of street (m)	10/6400	10/6,400
Routine maintenance: Seal @ 5 yrs (\$/m²)/ Seal @ 10 yrs (\$/m²)	2.6/2.6	2.6/2.6
Routine maintenance: Resurfacing 2" @10yrs (\$/m²)/ : Resurfacing 2" @20yrs (\$/m²)	8.5/8.5	na/8.5
Construction cost: CONC PVMT (JOINTED-PCP)(8") (\$)	3,858,351	3,858,351
Construction cost: Stabilization (material), (\$) + stabilization (labor), (\$)	0	318,618
Work zone/traffic data: AADT construction year (total for both directions)	2,000–4,000	
Work zone/traffic data: Single unit trucks/buses as percentage of AADT	8%	
Work zone/traffic data: Annual growth rate of traffic	4%	
Work zone/traffic: Lanes open in each direction under normal/ work zone conditions	1/ two way lane shift (½)	
Work zone/traffic data: Work zone duration (days)	20	27
Value of user time: Average value of time for passenger cars/trucks/buses	21 (\$/hour)	

(NS), (b) stabilisation (S) and (c) soil substitution (SUB) – were considered, as shown in Table 3.1. Medium-to-high truck traffic was considered (between 200,000 and 750,000 trucks/buses in the design traffic lane over the design life). It was estimated that the pavement life would be on an average 15 years for the control case (no stabilisation) and 20 years for the stabilisation as well as in the soil substitution (excavate and replace) cases.

3.3.2.2 Agency and user costs

In order to derive the user costs, the following parameters are taken into account:

- Work zone duration and length;
- Traffic data (AADT, lanes open in each direction, value of user time, traffic hourly distribution).

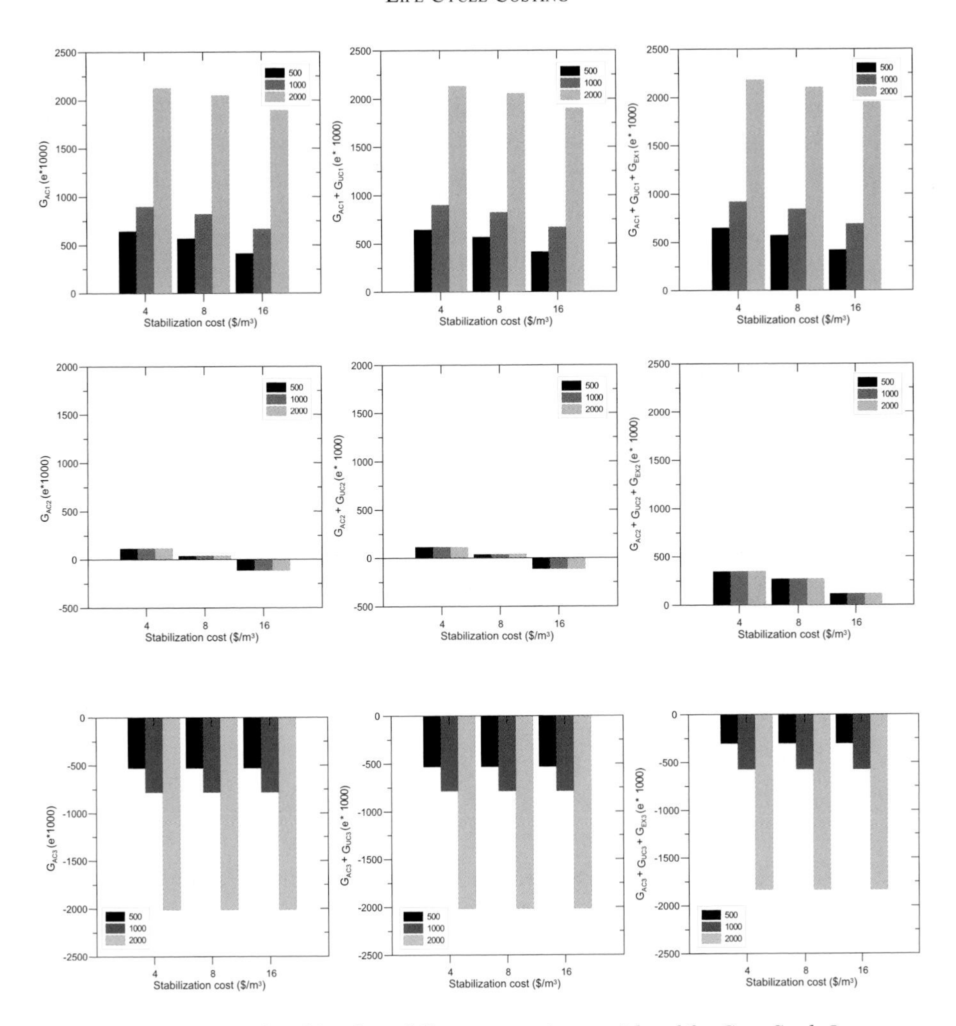

Figure 3.2 *Gains calculated for three different scenarios considered for Case Study I (Praticò et al., 2011b).*

It was possible to estimate the added time and vehicle stopping costs. Based on the aforementioned data, it was also possible to derive UC and AC present values for the cases under examination. Figure 3.2 summarises the results comparing the alternatives S vs. NS (i.e. $G_1 > 0$, if S is better than NS), S vs. SUB (i.e. $G_2 > 0$, if S is better than SUB) and NS vs. SUB (i.e. $G_3 > 0$, if NS is better than SUB). Comparing the agency costs, S usually gave better results than SUB (when the stabilisation cost was 4–8 €/m³), and this latter cost usually prevailed on NS. The G_{AC1} result was always positive, but the G_{AC2} was always negative when the stabilisation cost was substantially higher than the cost related to the removal and replacement option. This condition was for the cases under investigation, because the mechanical properties of very good soil were found to be substantially equivalent to that of stabilised soil. A key factor, from this standpoint, relied on the optimisation of the mechanical properties of stabilised soil, for the given cost of soil stabilisation. The higher the resilient modulus, the higher the probability of a longer expected life and a positive gain in stabilisation. In the case of the user cost consideration, due to the low-volume traffic conditions, PV_{UC} resulted in a lower value than the present value of agency costs (ratio 1: 50 c.a). The gains related to user and agency costs (e.g. $G_{AC} + G_{UC}$) followed the G_{AC} trends. This was because for LVRs, the user costs can be negligible (Ou and Swarthout, 1986). Stabilisation and substitution (SUB) resulted in substantially higher costs, while NS resulted in lower costs than for the remaining alternatives. It was noted that when the WZD due to soil stabilisation (S) or substitution (SUB) was higher than the control case (e.g. 1.5 times), then GUC1 approached negative values.

3.3.2.3 Environmental and total costs

Environmental gains G_{EX1}, G_{EX2} and G_{EX3} results were usually positive (option S was superior to NS; NS superior to SUB). It was also noted that when quarrying and landfill environmental costs increased and/or traffic decreased, G_{EX1} moved towards negative values and G_{EX2} and G_{EX3} moved towards positive values (i.e. NS performance was superior to S and S performance was superior to SUB). This fact was highly relevant because it implied that the creation of an environmental demand can strongly modify the ranking towards a clear prevalence of NS over SUB, especially for very low traffic volumes.

The overall gains ($G_{AC} + G_{UC} + G_{EX}$) were strongly affected by the cost of soil stabilisation (S), especially when comparing S vs. SUB (G_2): At lower cost, the more preferable way to stabilise would be with respect to removing and replacing. Total gains were usually positive for the comparisons 1 and 2 and negative for 3 (i.e. S better than SUB and SUB superior to NS). Very low traffic levels resulted in a tendency towards more sustainable pavement management systems, SUB becoming the inferior solution (S and NS were both superior to SUB).

Overall, based on the net present value (NPV) calculations, the lime-stabilised (S) subgrade sections gave lowered life cycle costs for large traffic volumes. However, for low traffic volumes, the substitution (SUB) option was uneconomical when compared to the no stabilisation (NS) option.

3.3.3 Case Study II – Arlington, Texas, United States

The construction of a low-volume local road (International Parkway) in Arlington, Texas, United States, was considered. Table 3.2 and Figure 3.3 summarise this case study. Prior to the pavement reconstruction, the previously existing asphalt layer had been subjected to severe longitudinal and transverse cracking, plus vertical movements due to the expansive subgrade soil. The City of Arlington decided to reconstruct the pavement structure by placing a concrete layer over a *combined lime–cement stabilised* subgrade. A series of laboratory tests were carried out on the natural and stabilised subgrade soils and the results are presented in Table 3.2. It was that the combined lime–cement stabilisation improved the mechanical properties of the stabilised soils. These laboratory experimental data were used in further LCC analysis. For LCC analysis purposes, three different stabilisation costs (Stb.cost ($\$/m^3$): 6, 12 and 25, as shown in Figure 3.3) and three traffic densities (1000, 2000 and 4000 vehicles/day, as shown in Figure 3.3) were considered. The same surcharge factors as used in case I were used to calculate the environmental costs.

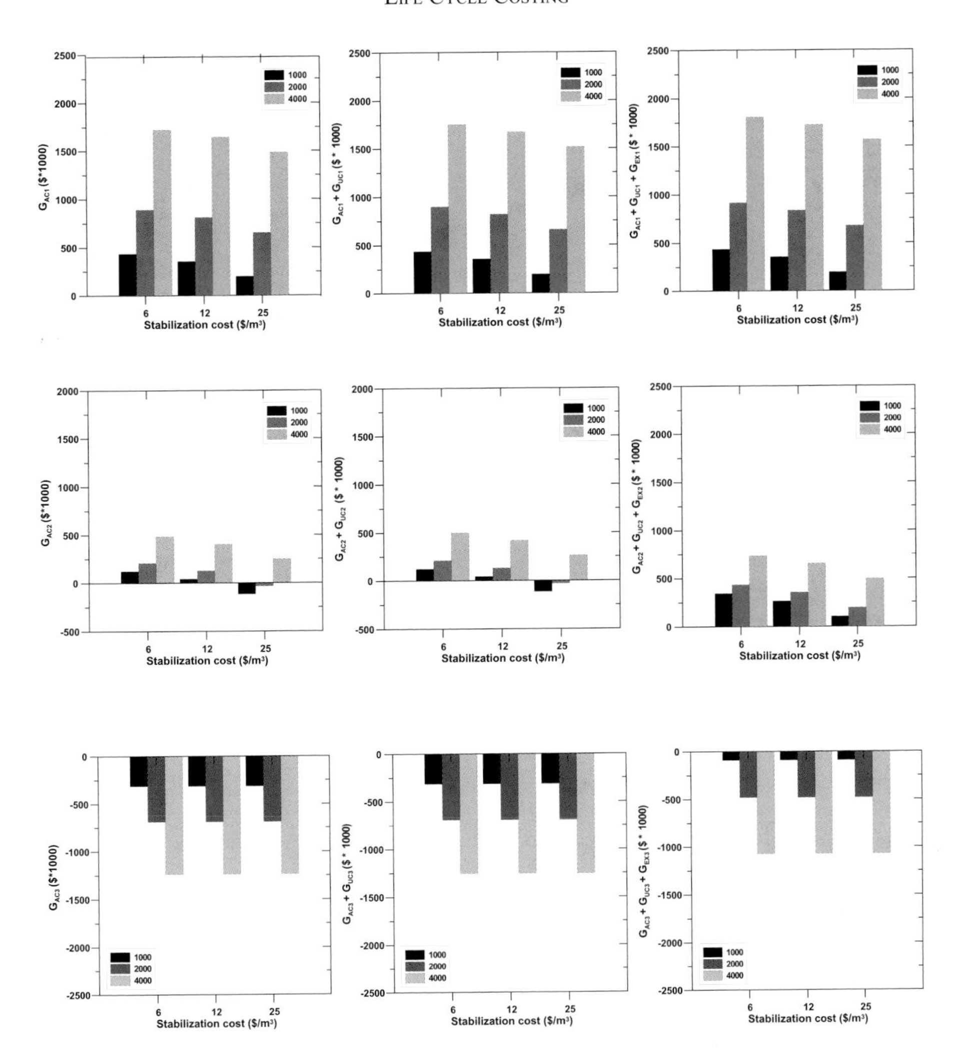

Figure 3.3 *Gains calculated for three different scenarios considered for Case Study II (Praticò et al., 2011b).*

3.3.3.1 Expected life

Construction costs were estimated from historical data provided by the City of Arlington (production rates, labour and equipment costs, material costs and maintenance costs). This information was used to estimate the unit prices for the control (NS) and the stabilisation (S) cases. The MEPDG and the AASHTO Guide 1993 were used to estimate the expected pavement life. A 75 per cent design reliability level was assumed in the analysis for the two-layered rigid pavement system. The mechanical characteristics of the Portland cement concrete (PCC) slab, and the other statistics required to estimate the expected life of the pavement are presented in Table 3.2. The environmental conditions corresponding to a southern United States climate were used. Three different design alternatives – (a) no stabilisation/control (NS), (b) stabilisation (S) and (c) soil substitution (SUB) – were considered. It was estimated that the average pavement life was 25 years for the control case NS (no stabilisation), 40 years for the stabilisation case (S) and 35 years for the soil substitution SUB (excavate and replace) case.

3.3.3.2 Agency and user costs

User costs were derived based on the parameters in Table 3.2. Since the LVR of interest is a two-lane/two-way road, only one lane was considered open for traffic during the maintenance and rehabilitation operations. Hence, the traffic flow in both directions was considered using the same traffic lane, while the other lane was being rehabilitated. Generally, a flagman or traffic signal was used to regulate the traffic in this scenario. In this case, the traffic delays were estimated to be higher. Figure 3.3 summarises the results comparing different alternatives – S vs. NS (i.e. $G_1 > 0$, if S is better than NS), S vs. SUB (i.e. $G_2 > 0$, if S is better than SUB) and NS vs. SUB (i.e. $G_3 > 0$, if NS is better than SUB).

If one compares the agency costs, S results were usually better than SUB and SUB was better than NS (when the stabilisation cost was reasonable, i.e. 6–12 \$/m^3). This gain also depended on the level of traffic. For lower traffic densities, the gain in agency costs was always negative (i.e. gain will be higher for higher traffic densities). This observation corroborates that the design alternative is dependent on traffic volume. G_{AC1} results were always positive; G_{AC2} is always negative when the stabilisation (S) costs were supposed to be substantially higher than the SUB cost. As described in the previous case study, the mechanical properties of a very good soil were found to be equivalent to that of a stabilised (S) soil. The higher the resilient modulus, the higher the probability of a longer expected pavement life and a positive gain for stabilisation (S). In the case of user cost considerations, due to the low-volume traffic conditions, user costs were lower than the agency costs. The stabilisation (S) and substitution (SUB) alternatives were almost equal, while NS results were lower than the remaining alternatives (S and SUB). When the WZD related to soil stabilisation (S) or substitution (SUB) is higher than that of the control case (e.g. 1.5 times), then $\mathrm{G}_{\mathrm{UC1}}$ approached negative values.

3.3.3.3 Environmental costs

Environmental gains G_{EX1}, G_{EX2} and G_{EX3} results are usually positive (S superior to NS; NS superior to SUB). When quarrying and landfill environmental costs increased and/or traffic decreased, G_{EX1} moved towards negative values and G_{EX2} and G_{EX3} moved towards positive values (i.e. NS performed better than S and S performed better than SUB). The overall gains ($G_{\mathrm{AC}} + G_{\mathrm{UC}} + G_{\mathrm{EX}}$) were strongly influenced by the cost of soil stabilisation (S), especially when comparing the alternatives S vs. SUB (G_2). The gain usually ensured positive values for cases 1 and 2 and negative values for case 3 (i.e. S was better than SUB and SUB was better than NS). Typically, a very low traffic level resulted in a more sustainable pavement management system, SUB becoming the inferior solution to S and NS.

Overall, based on the net present value (NPV) calculations for this case, the NPV of combined lime and cement-stabilised subgrade (S) sections indicated lower long-term costs compared with the non-stabilised subgrade. Although the costs of stabilisation were added to the initial construction cost in the treated sections, the long-term costs were estimated as much lower, as the results of the lower maintenance and rehabilitation activities required due to the better performance of the pavements over the design period.

3.3.4 Comparison of Case Studies I and II

Since both the case studies have similarities in pavement geometry, construction and maintenance costs, and traffic volume, the results could be compared (AADT = 1000 ~ 2000). The overall gains ($G_{AC} + G_{UC} + G_{EX}$) in both the cases were compared. It was seen that the expected life of the pavement on combined lime–cement treated subgrade is higher than the lime-stabilised subgrade. This is attributed to the improved mechanical properties of the combined lime–cement stabilised subgrade and this influenced the 'S vs. SUB' comparison. For a given traffic volume, the initial construction costs were higher for the combined lime–cement stabilisation (S) process. The user cost comparisons were marginal for both alternatives; however, the combined lime–cement treatment resulted in longer maintenance operation intervals. These facts revealed that the combined lime–cement treatment of subgrades gave higher net benefits (gain) when the traffic volume was higher for a given soil type and other traffic conditions. Case study I showed that for low traffic volumes, the substitution (SUB) option was uneconomical when compared with the no stabilisation (NS) option in lieu of sustainable pavement management.

3.4 Further developments and perspectives

The properties of pavements affect their overall cost, and when a characteristic of a pavement fails to meet expectations, this constitutes a quality assurance problem for the agency (Burati *et al.*, 2003; Praticò, 2007; Chen and Guo, 2011), and a pay adjustment (PA) needs to be estimated. A PA is the actual amount, either in dollars or in dollars per area, weight, or volume, that is to be added to or subtracted from the contractor's bid price or unit bid price (Burati *et al.*, 2003). The assessment of a quality characteristic depends on sampling plans, measurements of surface or bulk properties, plus a number of error sources that could be discussed (Praticò *et al.*, 2013). To solve this problem, several strategies are available, based on LCCA. Quality-related PAs (Hughes *et al.*, 2011) and performance-related specifications (Anderson *et al.*, 1991; Hoerner and Darter, 2000; Epps *et al.*, 2002; Hand *et al.*, 2004; Hwang *et al.*, 2009; Jeong and El-Basyouny, 2010; Fulgro Consultants Inc., 2011; Gedafa *et al.*, 2012) are available.

As discussed earlier, the use of LCCA to derive pay adjustments implies consideration of the entire life cycle. Furthermore, cost analysis can be affected by a number of performance factors and properties of the transportation infrastructures. These requirements call for a sound conceptual framework for innovative quality measures.

3.4.1 Soundness of the conceptual framework and 'fixed' points

The LCCA model proposed by Weed (2001) provides a most helpful algorithm to derive the pay adjustment (PA_{old}) and is as follows:

$$PA_{old} = C \cdot \left(R^D - R^E\right) \cdot \left(1 - R^O\right)^{-1} \tag{3.21}$$

In Eq. 3.21, PA_{old} is the pay adjustment (the subscript old is added), R is the ratio between (1+i) and (1 + r), where i is the inflation rate (typically 0.04) and r the interest rate (typically 0.08). The duration, D (e.g. 20 years) is the expected life of the as-designed life (AD), E (e.g. 15 years) is the expected life of the as-constructed pavement (AC), O is the time between two successive rehabilitations or resurfacings (typically 10 years). Furthermore, C is the cost of the pavement (€ or €/m^2) at time 0 and C_{RE} the cost of rehabilitation in the given year. C and C_{RE} can be identical.

In the aforementioned algorithm, PA/C does not tend to –1 (i.e. –100%) when E tends to zero. Hence, Eq. 3.22 was recast, based on a different conceptual framework (Praticò, 2011, 2013):

$$PA_{11} = C \cdot \left(R^D - R^E\right) \cdot \left(1 - R^D\right)^{-1} \tag{3.22}$$

In Eq. 3.22, if E tends to D, PA tends to 0, while if E tends to 0, PA tends to –C and the penalty equals (in absolute value) the cost. Rehabilitation or resurfacings are included in Eq. 3.21, but are not included in Eq. 3.22.

With respect to resurfacings, they are scheduled at O, D + O, 2D + O, and so on. In contrast, for AC pavements, they are performed at O, E + O, E + D + O, E + 2D + O, etc. Consequently, it gives the following equations (where $F = 1$ when $E > O$, while $F = 0$ if $E \leq O$ and $O < D$):

$$PA_1 = PA_{11} + PA_{12} = C \cdot \left(R^D - R^E\right) \cdot \left(1 - R^D\right)^{-1} + C_{RE} \cdot R^O \left[1 - F + \left(R^D - R^E\right) \cdot \left(1 - R^D\right)^{-1}\right] \tag{3.23}$$

where C_{RE} is the cost of rehabilitation and C is the construction cost. F is a function which assumes the values: $F = 1$ when $E > O$, $F = 0$ if $E \leq O$ and $O < D$ (this latter condition is a default hypothesis (see Praticò, 2011, 2013).

The denominators Eq. 3.23 differ from the denominator of PA_{old} (see Eq. 3.21, in which there is "O" instead of "D").

Figure 3.4 illustrates the dependence of PA/C on time (Eqs. 3.21 and 3.22) for different values of R. The case of $R = 1.06$ is unrealistic (interest rate lower than the inflation rate, $R > 1$), while the remaining cases refer to $R < 1$. Note that (1) for E ≤ D the model after Praticò (2011, 2013) yields values as expected in the range (–1, 0), while the other model yields values lower than zero, and sometimes lower than –1; (2) higher R values yield lower PA/C values for E < D (higher penalties) and higher PA/C values for E > D (bonuses), for both models.

Figures 3.5 and 3.6 illustrate the variation of PA/C over time with reference to Eqs. 3.22 and 3.23 (with RE).

Note that (1) in both the cases for E = 0 and E = D, the pay adjustment tends to –1 and 0, respectively. The results for these two 'fixed' points imply that the corresponding boundary conditions are satisfied; (2) the difference between the two curves depends on the ratio between C_{RE} and C. Under the given hypotheses ($R = 0.96$, $C_{RE}/C = 0.1$), only minor variations in the dominion of E close to 10 years are noted. This fact depends on the 'impulsive' function F, which can be modelled in terms of the 'if function'.

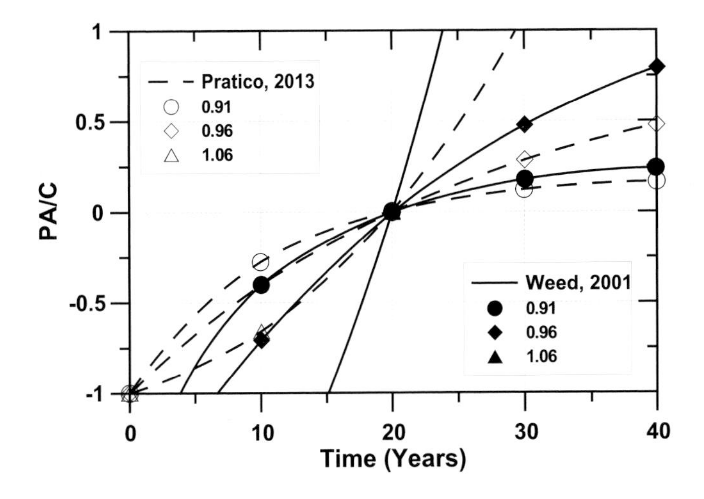

Figure 3.4 *Variation of PA/C with time.*

If both compulsory (e.g. friction and bearing characteristics) and premium properties (e.g. drainability and high albedo) are considered, based on the previous discussion, the following equation can be developed:

$$PA = PA_1(E_1) + \sum_{p=1}^{P} PA_p\left(\min\left(E_1, E_p\right)\right) \tag{3.24}$$

The overall PA is split into a compulsory ($PA_1(E_1)$) component and a premium component (the sum of PAs pertaining to premium characteristics). The subscript p refers to the pth premium performance characteristic. The expected life of the pavement, as a function of the pth premium performance characteristic, is E_P. The expected life as a function of the compulsory properties, E_1, is defined as follows:

$$E_1 = \min\left[E_M, \mathrm{E_F}, \ldots\right]. \tag{3.25}$$

where E_M is the expected structural life and E_F is the friction-based expected life.

3.4.2 Data variability and quality measures

A quality measure is a mathematical tool used to quantify the level of quality of an individual quality characteristic (Gharaibeh *et al.*, 2010; Seo, 2010; Hughes *et al.*, 2011; Praticò, 2013). Examples of quality measures are mean, standard deviation, per cent defective (PD), per cent within limits (PWL), average absolute deviation (AAD), moving average, and conformal index (CI). PWL and PD are the quality measures typically recommended for use in quality assurance specifications. Per cent defective is the percentage of a sampling lot falling outside the specification limits (to the left of LSL-lower specification limit, or to the right of USL, upper specification limit).

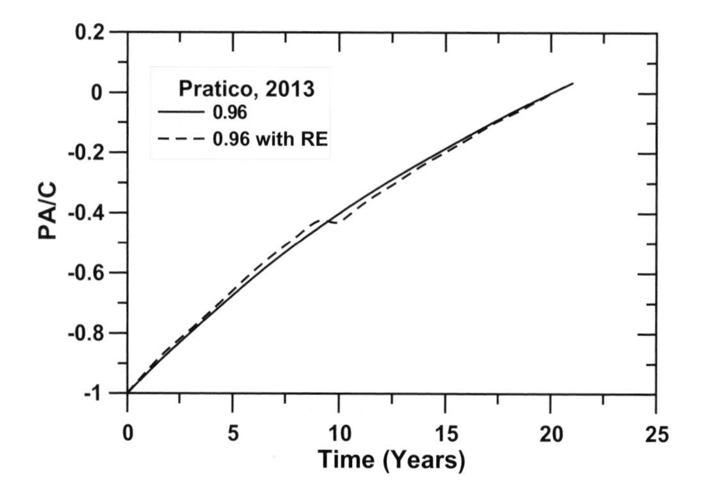

Figure 3.5 *Variation of PA/C with time (boundary conditions for E = 0 and E = D).*

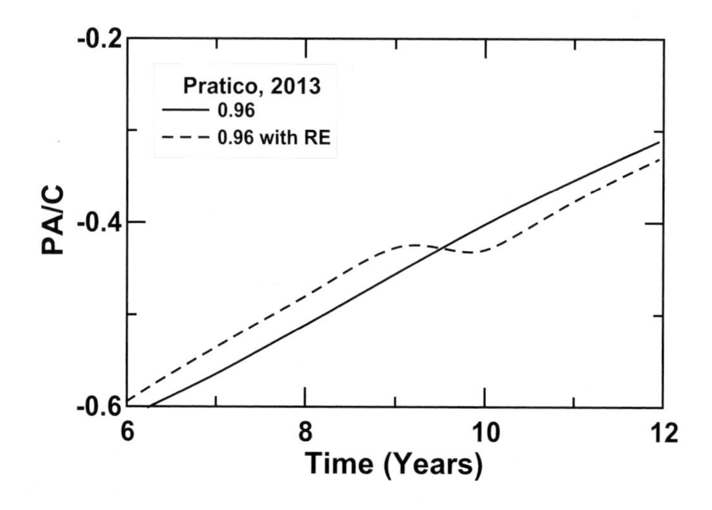

Figure 3.6 *Variation of PA/C with time: local effect of resurfacings (dotted line).*

PWL and PD (PD = 100-PWL) are considered the most effective measures of both central tendency and variability and are often used as quality measures by US State highway agencies, the US Federal Aviation Administration and the US Federal Highway Administration (Hughes *et al.*, 2011). Furthermore, they are used to derive performance models, where the expected life of the pavement (independent variable) is related to the PDs of several hot-mix asphalt characteristics (air void content, AV; thickness, TH, etc).

Note that for both PD and PWD, a datum with given value $x + 0.5(\text{LSL} + \text{USL})$, that is higher than the average, has the same effect as a datum with value $0.5(\text{LSL} + \text{USL}) - x$, that is lower than the average.

This property makes PD and PWL quite inconsistent for use in the aforementioned performance models. For example an excess of thickness would cause a very different effect compared to a lack of thickness, and the consequences of an excess versus a deficiency in air voids would be completely different.

Furthermore, data variability and normality are considered the main issues to be addressed (Katicha *et al.*, 2011) and several drawbacks concerning PWL effectiveness have been highlighted in dealing with processes strongly affected by central tendency (Praticò, 2011; Praticò *et al.*, 2012).

For this reason, the indicator PD* was introduced (Praticò, 2013). PD* is a PD with a positive or negative sign, depending on the relationship between the average of the distribution and the mid-point between LSL (the lower specification limit) and USL (the upper specification limit). For example, if the variable is AV (and AVm is its mean), then PD* is defined as follows:

$$PD^* = PD \text{ if } AVm \geq 0.5 \cdot (\text{LSL} + \text{USL}) \tag{3.26}$$

$$PD^* = -PD \text{ if } AVm < 0.5 \cdot (\text{LSL} + \text{USL}) \tag{3.27}$$

Another method to deal with data variability without losing data 'directionality', that is for considering both position and dispersion effects, was proposed by Praticò *et al.* (2012). This method is based on the following assumptions: (1) the traditional LCCA-based method (equation n) is partly inconsistent because it does not depend on the dispersion (variability) of the quality characteristics; (2) the PD-based methods are intrinsically empirical due to lack of well-grounded links between PA and the expected life; consequently (3) an excess of variability in the layer quality characteristics would not provide variable pay adjustment in the first class of methods because the averages of the main quality characteristics are unchanged. Similarly, pavement with a higher expected life can correspond to a lower pay factor in the second class of methods. In summary, both classes of methods can lead to unreasonable conclusions by following their implications logically to absurd and inconsistent consequences.

Based on the previous assumptions, the following generalised algorithm was proposed:

$$PA^* = PA^*_{\text{LCCA}} + F^*(\sigma_{AV}, \sigma_{TH}, \ldots), \tag{3.28}$$

where PA* is the PA-to-C ratio and PA^*_{LCCA} can be derived according to equations of type 3.21 or 3.22. F^* is a function and σ_{AV}, σt are standard deviations of the quality characteristics that can affect the expected life of the as-constructed pavement.

These equations are intrinsically related to nonlinearity of the E vs. AV relationship and are consistent with (1) the need for homogeneous production and (2) an increase in the life cycle cost for increasing the number of work zones (agency costs, etc.). Furthermore, the following simplified equation is proposed:

$$PA^* = PA^*_{\mathrm{LCCA}} - k \cdot \sigma_{\mathrm{AV}}, \tag{3.29}$$

where k ranges from 2.3 to 4.7 and only accounts for an increase in the absolute value of the PA due to the peculiar assumptions under investigation (the normality of AV and the derivation of E through the M-EPDG).

3.5 Conclusions

In this chapter, a new conceptual engineering-economic tool based on life cycle cost analysis is discussed to optimise and select stabilisation alternatives for a better performance of a pavement, for given subgrade soils and traffic conditions. The agency, user and environmental costs are separately and synergistically addressed and incorporated in the new model. The model was thoroughly calibrated and validated with two case studies having similarities in geometry, construction costs and maintenance costs, and traffic volume, however, with distinct geographical conditions. As a part of this study, a simple Excel spreadsheet was used for the model implementation and calibration.

It is evident from the analyses that the estimation of both user and environmental costs is critical and care must be taken while estimating these costs. The agency costs are easy to obtain and reliable in the analysis. The effects of stabilisation (S) and the use of alternative construction materials must be evaluated for both the laboratory and field conditions before considering them in the analysis. The corresponding environmental costs may be considered in the analysis with appropriate simplifications and generalisations, as discussed in this chapter.

It was observed that the actual gains are dependent on the traffic volume and the incurred stabilisation costs. The user and environmental values tend to modify the ranking in favour of soil stabilisation (S) and the zero hypotheses (NS) if the traffic level is high. For very low traffic, the no stabilisation (NS) (control) option with a perpetual pavement system results in higher cost benefits.

Based on the gain and NPV calculations, the gain of combined lime–cement stabilised (S) subgrade sections have indicated lower long-term costs compared to non-stabilised subgrade (NS) or lime stabilised (S) subgrade sections. Although the cost of stabilisation was added to the initial construction costs in the treated sections, the long-term maintenance costs were lower due to the results of lower maintenance activities in the treated sections because of the better performance of the pavements.

Finally, in comparison of case studies I and II, for given traffic volumes and site conditions, the net gains are high for higher traffic volumes in the case of combined lime–cement treatment than when lime-treated subgrade soils are involved. Overall, the newly developed LCCA model can be used successfully to evaluate different stabilisation alternatives considered for increased performance of pavements.

Nomenclature

AC	Agency cost, $	NLO	Number of lanes open
C	Initial construction cost, $	O	Expected life of successive rehabilitations, years
CA_{ki}	Truck capacity, m^3	PV_{AC}	Present value of agency cost, $
C_{max}	Maximum initial construction cost, $	PV_D	Present value due to delay, $
D_{kj}	Average distance to cover, km	PV_{UC}	Present value of user cost, $
E	Expected life, year	PV_{VOC}	Present value of vehicle operating costs, $
EM_{kj}	Emission surcharge factor, $/l	R_k	Rehabilitation cost, $
EX	Environmental cost, $	R_{kmax}	Maximum rehabilitation cost, $
EX_0	Environmental cost during construction, $	S_a	Salvage value, $
EX_{AD}	Additional environmental cost, $	SC	Stopping cost, $
EX_{EM}	Environmental cost due to emission, $	T	Period of analysis, year
EX_k	Environmental cost during the kth rehabilitation, $	TD	Traffic data
EX_{LA}	Environmental cost related to landfilling, $	TOD	Traffic hourly distribution
EX_{QU}	Environmental cost related to quarrying, $	UC	User cost, $
FC_{kj}	Fuel consumption, l/km	VN_{kj}	Volume of material needed, m^3
G	Gain, $	VUC_k	Value of user costs, $/hr
G_{AC1}	(PVAC)NS – (PVAC)S, $, gain in agency cost for choosing S instead of NS	WZC	Work zone capacity, vpd
G_e	Pavement geometry	WZD_k	Work zone duration, hr
I	Interest rate, %	WZL_k	Work zone length, m
M	Mechanical properties	WZSL	Work zone speed limit, kmph
n_k	Expected life of *k*th rehabilitation, years	ΔT	Added time, hr
$n_1...n_N$	Expected lives, years		

References

AASHTO (American Association of State Highway and Transportation Officials) (1993) *Guide for design of pavement structures*. Washington, D.C.

AbouRizk, S.O.S. and S. Ariaratnam (2003) Risk-based life-cycle costing of infrastructure: rehabilitation and construction alternatives, *Journal of Infrastructure Systems*, **9**, 6–15.

Anderson, D.A., D.W. Christensen and H. Bahia (1991) Physical properties of asphalt cement and the development of performance-related specifications. *Journal of the Association of Asphalt Paving Technologists*, **60**, 437–475.

Basma, A.A. and E.R. Tuncer (1991) Effect of lime on volume change and compressibility of expansive clays. *Transportation Research Record: Journal of the Transportation Research Board,* **1296**, 54–61 (Transportation Research Board of the National Academies, Washington, D.C.).

Behrens, I.L.C. (1999) Overview of low-volume roads. *Transportation Research Record: Journal of the Transportation Research Board,* **1652**, 1–4. (Transportation Research Board of the National Academies, Washington, D.C.).

Bugge, W.A. and R.R. Bartelsmeyer (1961) Soil stabilization with Portland cement. *Highway Research Board, 292, National Research Council*, Washington, D.C., 1–15.

Burati, J.L., R.M. Weed, C.S. Hughes and H.S. Hill (2003) Optimal procedures for quality assurance specifications. *Office of Research, Development, and Technology, Report FHWARD-02-095*, FHWA McLean VA, 347.

Bushman, W.H., T.E. Freeman and E.J. Hoppe (2005) Stabilization techniques for unpaved roads. *Transportation Research Record: Journal of the Transportation Research Board,* **1936**, 28–33 (Transportation Research Board of the National Academies, Washington, D.C.).

Chen, Y. and D. Guo (2011) Study on payment methods of expressway based on project quality, *Transportation, Mechanical, and Electrical Engineering (TMEE)*, 2011 International Conference.

Chesner, W.H., M.J. Simon. and T.T. Eighmy (2001) Recent federal initiatives for recycled materials use in highway construction in the United States. *Beneficial Use of Recycled Materials in Transportation Applications*, November 13–15, 3–10.

Croft, J.B. (1967) The structures of soils stabilized with cementitious agents. *Engineering Geology*, **2**, 63–80.

Das, B.M. (1988) *Principles of geotechnical engineering*. PWS Publishing Company Boston.

Douglas, D.G. and R.K. Molenaar (2004) Life-cycle cost award algorithms for design/build highway pavement projects. *Journal of Infrastructure Systems*, **10** (4).

Ellis, S.J. (2003) Recycling in transportation. In *Transportation geotechnics symposium* (Frost M. W., I. Jefferson, E. Faragher, E.J. Rofft, and P.R. Fleming (eds)). Thomas Telford, London, 177–188.

Epps, J.A., A.J. Hand, A.J., S. Seeds, T. Schotz, S. Alavi, C. Ashmore, C.L. Monismith, J.A. Deacon, J.T. Harvey and R. Leahy (2002) Recommended performance-related specification for hot-mix asphalt construction: results of WesTrack project. *NCHRP Report No. **455**, Transportation Research Board.*

FHWA (2002) *Life-cycle cost analysis primer 2002.* Office of the Asset Management, United States Department of Transportation, 25.

Fulgro Consultants Inc. (2011) A performance-related specification for hot-mixed asphalt, NCHRP Report 704, *Transportation Research Board.*

Gedafa, D.S., M. Hossain, L.S. Ingram and R. Kreider (2012) Performance-related specifications for PCC pavements in Kansas. *Journal of Materials in Civil Engineering,* **24(4),** 479–487.

Gerbrandt, R. and C. Berthelot (2007) Life-cycle economic evaluation of alternative road construction methods on low-volume roads. *Transportation Research Record: Journal of the Transportation Research Board*, **1989**, Transportation Research Board of the National Academies, Washington, D.C., 61–71.

Gharaibeh, N.G., S. Garber and L. Liu (2010) Determining optimum sample size for percent-within-limits specifications. *Transportation Research Record: Journal of the Transportation Research Board*, **2151**, 77–83.

Hall K.D. and J.W. Bettis (2000) *Development of comprehensive low-volume pavement design procedures*, Final Report, Report No. MBTC 1070, *Arkansas State Highway and Transportation Department.*

Hand, A.J., A.E. Martin, P.E. Sebaaly and D. Weitzel (2004) Evaluating field performance: case study including hot mix asphalt performance-related specifications. *Journal of Transportation Engineering*, **130(2).**

Hoerner, T.E. and M.I. Darter (2000) Improved prediction models for pcc pavement performance-related specifications. *Volume I: Final Report, FHWA*, September 2000. http://www.ohio.edu/icpp/upload/*A Practical Guide to Low-Volume Road-Newcomb*.pdf accessed 01/09/14.

Hughes, C.S., J.S. Moulthrop, S. Tayabji, R.M. Weed and J.L. Burati (2011) Guidelines for quality-related pay adjustment factors for pavements. *NCHRP Project No. 10-79*, Final Report, 2011 – onlinepubs.trb.org.

Hwang, S.-M., S.-K. Rhee and S.-M. Kim (2009) Establishment of performance related specifications using pay factors and relationship between fatigue cracking and pay factors. *Material, Design, Construction, Maintenance, and Testing of Pavement: Selected Papers* from the 2009 *GeoHuman International Conference.*

Ian, W.H., I.W.H. Parry and G.R. Timilsina (2009) Pricing externalities from passenger transportation in Mexico City, Policy Research Working Paper 5071. *The World Bank, Development Research Group – Environment and Energy Team*, September 2009.

Jeong, M.J. and M. El-Basyouny (2010) Statistical applications and stochastic analysis for performance-related specification of asphalt quality assurance, transportation research record. *Journal of the Transportation Research Board*, **2151**, 84–92.

Katicha, S.W., G.W. Flintsch, K. McGhee and E. De León Izeppi (2011) Variability and normality assumptions for Virginia Department of Transportation volumetric properties: analysis of contractor data. *Transportation Research Record Journal of the Transportation Research Board, Washington, D.C.*

Keller, G. and J. Sherar (2003) *Low-volume roads engineering – best management practices field guide.* USDA, Forest Service Washington, D.C.

Kim, J.R. and D.E. Newcomb (1991) *Low volume road pavement design: a review of practice in the Upper Midwest*, University of Minnesota, MPC Report No. 91–3.

Kota, B.V.S. (1996) Sulfate bearing soils: problems with calcium based stabilizers. *Transportation Research Record: Journal of the Transportation Research Board,* **1546**, Transportation Research Board of the National Academies, Washington, D.C, 1996, 62–69.

Little, N.D. (1995) *Stabilization of pavement subgrades and base courses with lime.* Kendall/Hunt Publishing Co., Dubuque, IA.

Maryvonne, F.P. (2007) Planning roads for rural communities. In *Transportation Research Record: Journal of the Transportation Research Board*, **1989**, Transportation Research Board of the National Academies, Washington, D.C., 1–8.

MEPDG (2007) *Guide for mechanistic-empirical design of new and rehabilitated pavement structures.* www.trb.org/mepdg. National Cooperative Highway Research Program, Transportation Research Board, Washington, D.C.

Muench, S.T., G.C. White, J.P. Mahoney, L.M. Pierce and N. Sivaneswaren (2004) Long-lasting low-volume pavements in Washington State. International Symposium on Design and Construction of Long Lasting Asphalt Pavements. *International Society for Asphalt Pavements*, Auburn, AL, 729 –773.

Nunes, M., M. Bridges and A. Dawson (1996) Assessment of secondary materials for pavement construction: technical and environmental aspects. *Waste Manage*, **16**, pp. 87–96.

Olof, J. (1997) Optimal road-pricing: simultaneous treatment of time losses, increased fuel consumption, and emissions. *Transportation Research Part D: Transport and Environment*, **2** (2), 77–87.

Ou, F.L. and C.D. Swarthout (1986) Cost-estimating model for low-volume roads. *Transportation Research Record: Journal of the Transportation Research Board*, **1055**, Transportation Research Board of the National Academies, Washington, D.C, 51–56.

Praticò, F.G. (2004) A theoretical and experimental study of the effects on mixes added with RAP caused by superpave restricted zone violation. *Journal of Road Materials and Pavement Design*, **5**(1), 73–91.

Praticò, F.G. (2007) Quality and timeliness in highway construction contracts: a new acceptance model based on both mechanical and surface performance of flexible pavements. *Construction Management and Economics*, **25**(3), 305–313.

Praticò, F.G. (2011) Pay adjustment in construction engineering. *International Conference on Business Intelligence and Financial Engineering*, December 12–13, 2011, Hong Kong.

Praticò, F.G. (2013). New road surfaces: logical bases for simple quality-related pay adjustments. *Journal of Construction Engineering and Management*, **139**(11), 04013020.

Praticò, F.G., R. Ammendola and A. Moro (2010) Factors affecting the environmental impact of pavement wear. *Transportation Research Part D: Transport and Environment*, **15**(3), 127–133.

Praticò, F.G., A. Casciano and D. Tramontana (2011a) Pavement life cycle cost and asphalt binder quality: a theoretical and experimental investigation. *Journal of Construction Engineering and Management*, **137**(2), 99–107.

Praticò, F.G., S. Saride and A. Puppala (2011b) Comprehensive life-cycle cost analysis for selection of stabilization alternatives for better performance of low-volume roads. *Transportation Research Record*, **2204**, 120–129.

Praticò, F.G., A. Casciano and D. Tramontana (2012) The influence of dispersion and location on pay adjustment in construction engineering. *Journal of Construction Engineering and Management*, **138**(10), 1125–1130.

Praticò, F.G., R. Vaiana, and A. Moro (2013) The dependence of volumetric parameters of hot mix asphalts on testing methods. *Journal of Materials in Civil Engineering*, 10.1061/(ASCE)MT.1943-5533.0000802 (Feb. 15, 2013).

Puppala, A.J., C. Dunder, S. Hanchanloet and K. Ghanma (1998) Swell and shrinkage characteristics of lime-treated sulfate soils. *Texas ASCE Spring Meeting Proceedings*, South Padre Island, 1–10.

Puppala, A.J., E. Wattanasanticharoen and L.R. Hoyos (2003) Ranking of four chemical and mechanical stabilization methods to treat low-volume road subgrades in Texas. *Transportation Research Record: Journal of the Transportation Research Board*, No. **1819**, Transportation Research Board of the National Academies, Washington, D.C, 63–71.

Puppala, A.J., G.S. Pillappa, L.R. Hoyos, D. Vasudev and D. Devulapalli (2007) Comprehensive field studies to address the performance of stabilized expansive clays. In *Transportation Research Record: Journal of the Transportation Research Board*, No. **1989**, Transportation Research Board of the National Academies, Washington, D.C, 2007, **2**, 3–12.

Saride, S., A.J. Puppala and R. Williammee (2010) Assessing recycled/secondary materials as pavement bases, special issue on sustainability in ground improvement projects. *Proceedings of ICE, Ground Improvement*, **163**(1), 3–12.

Seo, Y. (2010) Development and implementation of Korea's first percent within limit (PWL) specification for road pavements. *KSCE Journal of Civil Engineering*.

Sherwood, P.T. (1993) *Soil stabilization with cement and lime*. Transport Research Laboratory, Department of Transport, London.

Sirivitmaitrie, C. (2008) *Novel stabilization methods for sulfate and non-sulfate soils*. PhD Thesis, The University of Texas at Arlington. p. 294.

Timm D.H., D.E. Newcomb and S. Immanuel (2006) A practical guide to low-volume road perpetual pavement design. *International Conference on Perpetual Pavement Ohio Research Institute for Transportation and the Environment*, May 26.

Walls, J. and M.R. Smith (1998) Life-cycle cost analysis in pavement design, *Interim Technical Bulletin*, Federal Highway Administration.

Weed, R.M. (2001) *Derivation of equation for cost of premature pavement failure*. Transportation Research Record, **1761**, Transportation Research Board, Washington, D.C., 2001.

Wilde, W.J., S. Waalkes and R. Harrison (1999) *Life cycle cost analysis of Portland cement concrete pavements*, Report No. FHWA/TX-00/0-1739-1, U.S. Department of Transportation.

Willis, K.G. and G.D. Garrod (1999) Externalities from extraction of aggregates: regulation by tax or land-use controls. *Resources Policy*, **25**(2), 77–86.

WSDOT (1995) *WSDOT pavement guide*. Washington State Department of Transportation, Olympia, WA.

Yin, Y. and L. Siriphong (2006) Internalizing emission externality on road networks. *Transportation Research Part D: Transport and Environment*, **11**(4), 292–301.

Chapter 4
Pavement type selection for highway rehabilitation based on a life-cycle cost analysis: validation of California interstate 710 project (Phase 1)

Eul-Bum (E.B.) Lee, Changmo Kim and John T. Harvey

Summary

Life-cycle cost analysis (LCCA) for highway projects is an analytical technique that uses economic principles in order to evaluate long-term alternative investment options, especially for comparing the value of alternative pavement structures and strategies. Recently, the California Department of Transportation (Caltrans) mandated LCCA implementation to evaluate the cost effectiveness of pavement design alternatives for highway projects in the state. In this chapter, an LCCA approach was utilised for the validation of the pavement design on the I-710 Long Beach rehabilitation project with three alternative pavement types: (1) innovative (long-life) asphalt concrete pavement (ACP), (2) standard-life ACP, and (3) long-life Portland cement concrete pavement (PCCP). The LCCA followed Caltrans procedure and incorporated information filed by the project team. The software tools *CA4PRS* (Construction Analysis for Pavement Rehabilitation Strategies) and *RealCost* (USDOT, 2002) were used for quantitative estimates of construction schedule, work zone user cost, and agency cost for initial and future maintenance and rehabilitation activities. Conclusions from the LCCA supported use of the innovative ACP alternative, the one actually implemented in the I-710 Long Beach project (Phase 1), as it had the lowest life-cycle costs over the 60-year analysis period. For example, the life-cycle agency cost for the innovative ACP alternative ($33.2 million) was about $7.9 million more cost-effective than the Standard ACP alternative ($41.1 million) and about $17.2 million less expensive than the long-life PCCP alternative ($50.4 million). Utilization of the proposed computer tool-aided LCCA procedure would contribute substantial economic benefits to nationwide highway projects, especially rehabilitation and reconstruction.

4.1 Introduction

Life-cycle cost analysis (LCCA) is an analytical technique that uses economic principles to evaluate long-term alternative investment options in highway construction.

LCCA accounts for costs relevant to the sponsoring agency, owner, facility operator, and roadway user that will occur throughout the life of an alternative. Relevant costs include those for initial construction, future maintenance and rehabilitation (M&R), and user costs (time delay and vehicle operation costs in the work zone). The LCCA analytical process helps to identify the lowest cost-alternative that will accomplish a highway project's objectives by providing critical information for the overall decision-making process. For the last decade, LCCA has been emphasised as much as the initial project cost analysis in evaluating the design and construction plans for highway projects.

Recently, the California Department of Transportation (Caltrans) mandated implementation of LCCA in order to evaluate the cost effectiveness of alternative pavement designs for new roadways and existing roadways that require Capital Preventive Maintenance (CAPM), rehabilitation, or reconstruction (Caltrans, 2007a). The Caltrans *Highway Design Manual* (HDM) Topics 612 and 619 identify situations where an LCCA must be performed in order to assist in determining the most appropriate alternative for a project (Caltrans, 2007b). Since the cost impacts of a project's lifecycle are fully taken into account when making project-level decisions for pavements, Caltrans' practice is to perform an LCCA when scoping a project (Project Initiation Document phase).

Many researchers and practitioners have been developing LCCA concepts and computer tools to undertake the most efficient cost comparison of alternatives. Papagiannakis and Delwar developed a computer model to perform LCCA of roadway pavement, analysing both agency and user costs. Their software accepts inputs from a pavement management database and carries out pavement LCCA on both network-wide and project-specific levels (Papagiannakis and Delwar, 2001). Rather than considering user delay and future M&R costs, this software calculates the net annualised savings in user costs as the benefit that results from reducing pavement roughness (e.g. vehicle depreciation, maintenance, and repair, tires, and cargo damage) from its current condition to that in the end year of the life-cycle.

Salem *et al.* introduced a risk-based probabilistic approach to predict probabilities of the alternative occurrence of different life-cycle costs on infrastructure construction and rehabilitation. Their model predicts probability of time of infrastructure failure for building alternatives (Salem *et al.*, 2003). Using the Florida and Washington State Department of Transportation project databases, Gransberg and Molenaar developed best-value award algorithms of life-cycle cost for design/build highway pavement projects (Gransberg and Molenaar, 2004). Labi and Sinha studied the cost effectiveness of different levels of life-cycle preventive maintenance (PM) for three asphalt concrete (AC) functional class families and presented a methodology to determine optimum PM funding levels based on maximum pavement life (Labi and Sinha, 2005).

In 2002, the Federal Highway Administration (FHWA) first published an LCCA primer to provide sufficient background and demonstrations for transportation officials (FHWA, 2002, 2004). In addition, in 2004 the FHWA distributed an LCCA software tool, *RealCost (Version 2.1)*, to support practitioners performing LCCA for highway projects (FHWA, 2004). The Caltrans Office of Pavement Management and the University of California Pavement Research Center (UCPRC) have enhanced *RealCost* software and customised it for California (1) to add to its analytical capability for cost estimation, (2) to improve work zone traffic analysis, and (3) to implement automatic future M&R sequencing.

4.2 I-710 Long Beach rehabilitation project (Phase 1)

Interstate 710 (I-710), known as the Long Beach Freeway, opened in 1952 and serves as a major route for commuter and commercial traffic between Los Angeles and Long Beach (Figure 4.1). It is also a gateway to the ports of Long Beach and Los Angeles, two of the busiest cargo ports in the United States. However, a 51.5 km stretch of the I-710 corridor had become seriously deteriorated and required rehabilitation to keep it safe for road users. Caltrans consequently devised a rehabilitation project for I-710 and divided the plan into four phases. To date, Phases 1 (2003) and 2 (2010) have been completed, while Phases 3 and 4 are scheduled for the near future. Given the need for minimal disruption of heavy weekday traffic, Caltrans decided to carry out the pavement rehabilitation on I-710 with 55-hour extended closures, Caltrans' typical Long-Life Pavement Rehabilitation Strategy (LLPRS) practice for urban corridor networks.

The scope of the I-710 Long Beach project (Phase 1) was to rehabilitate the approximately 4.4 centreline-km of existing concrete pavement on I-710 near the city of Long Beach with long-life (30-year design) AC pavement (Figure 4.2) (Lee *et al.*, 2005). The existing pavement consisted of 200 mm (8 in.) Portland cement concrete

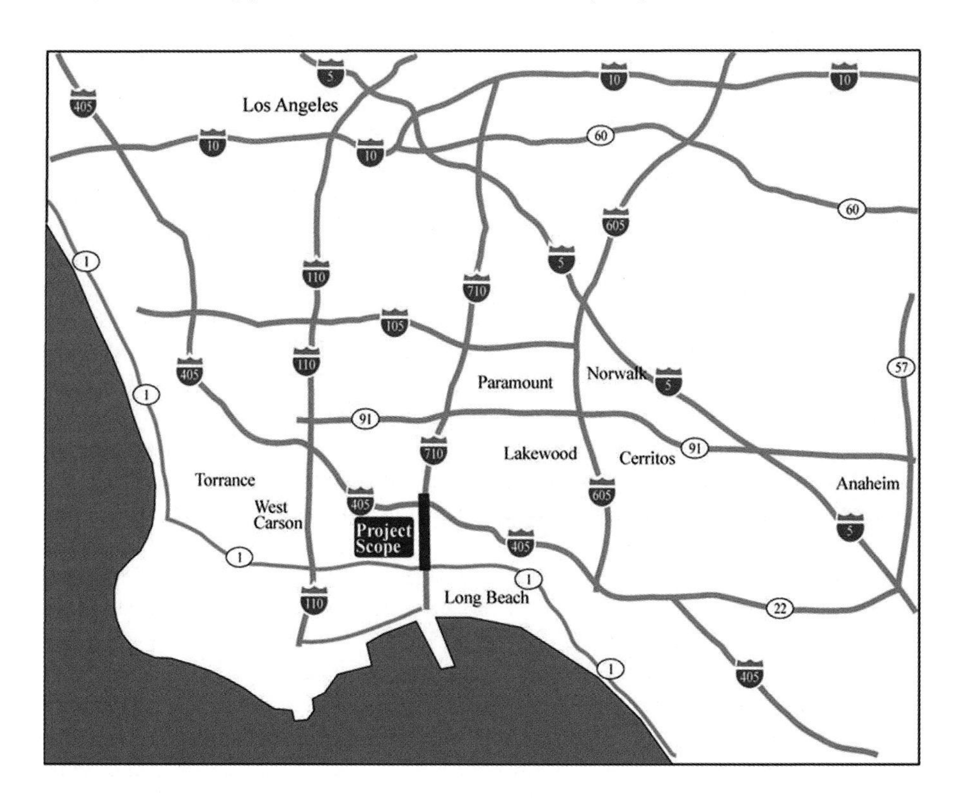

Figure 4.1 *Location of the California I-710 rehabilitation project (Phase 1: Long Beach).*

Figure 4.2 *Long-life AC pavement rehabilitation on I-710 with 55-hour extended weekend closure.*

(PCC) on top of 100 mm (4 in.) of cement-treated base, which was formerly Caltrans' most common rigid pavement type in the 1960s through 1970s. The rehabilitation included the three main lanes, the median, and the shoulders in each direction. Beneath the highway overpasses, which did not meet current federal bridge clearance requirements, the existing concrete pavement structure was removed – with an additional 150 mm (6 in.) excavation – and replaced with a total of 330 mm (13 in.) of new AC with five layers. Between the overpasses, the old concrete slab was cracked, seated (rolled), and overlaid with 230 mm (9 in.) of a new AC with four layers. Typically, two sections – a 400-m (1,300 feet) section of full-depth asphalt concrete (FDAC) replacement under the overpass, and a 1,200-m (4,000 feet) section of crack, seat, and overlay (CSOL) between the overpasses – were finished within one 55-hour closure.

Caltrans initially planned ten extended weekend closures with an incentive bonus, and the contractor successfully completed the rehabilitation with eight closures and received a $500,000 incentive bonus.

During the pavement rehabilitation with 55-hour extended weekend closures, Caltrans applied a counter-flow traffic system, completely closing off one side of the highway for construction and diverting traffic to the roadbed on the other side of the construction site through median crossovers. Using moveable concrete barriers, the outside shoulder was temporarily converted to a main traffic lane to provide two lanes in each direction. The main rehabilitation operations were performed with around-the-clock construction (nonstop) during 55-hour extended weekend closures (from 10 p.m. Friday to 5 a.m. Monday) to avoid weekday commute traffic.

A postconstruction summary report was recently published that included some periodic measurements of long-life AC pavement performance on I-710 Long Beach (Phase 1), including those made with the falling weight deflectometer at approximate annual intervals as well as some pavement noise and skid measurements, from the opening to traffic in summer 2003 to January 2009 (Monismith *et al.*, 2009). In summary, the overall actual performance of the I-710 Long Beach project measured for the last five years shows the long-life AC pavement behaving as it was designed to do. In addition, the rutting performance of the I-710 long-life AC pavement, which was measured in summer 2009 after 6 years of service traffic, shows a very positive indication on rut depth: approximately 5 to 6 mm (about 2 in.) at its greatest (roughly half of the expected long-life AC mix design criterion).

4.3 Study objective and analysis procedures

The primary objective of the LCCA study summarised in this chapter was to validate the benefit derived from selecting a long-life (for 30+ years design life) AC pavement over a 60-year analysis period. Following the formal Caltrans LCCA procedure, agency costs (including initial construction and maintenance costs) and road user costs (RUCs) (traffic delay) for the work zone are estimated for each alternative, utilizing the pavement engineering software tools *RealCost* and *CA4PRS*. In this postconstruction analysis, the baseline pavement type selected for the I-710 project (i.e. long-life AC pavement) is compared with other candidate alternative types from the perspective of savings of life-cycle agency cost and user cost.

The Caltrans LCCA procedures summarised below – and as specified in the department's procedure manual (Caltrans, 2007c) – have been applied to the validation of pavement type selection for the I-710 Long Beach rehabilitation project.

- Select several pavement design alternatives for initial construction, including pavement types, cross sections, materials, and expected design lives.
- Determine the analysis period to cover the design lives of the initial construction and future M&R activities. Caltrans recommends using a 60-year analysis period when long-life (30+ years) pavement design is compared in the LCCA.
- Identify future M&R activities for each design alterative, including pavement cross-section changes, sequence, and timeline over the analysis period.
- Analyse the schedule for the initial construction and future M&R activities.
- Estimate the project cost associated with the initial construction and future M&R.
- Calculate the RUC in the work zone for the initial construction and future M&R activities.
- Calculate the life-cycle cost for each design alternative using the concept of net present value (NPV) based on the discounted rate.
- Evaluate the LCCA results for the pavement design alternatives in terms of the benefits of life-cycle cost savings to validate and justify the adoption of long-life AC pavement for the I-710 project.

4.4 Engineering tools utilised

4.4.1 *CA4PRS* for schedule, traffic and cost analysis

This LCCA study utilised *CA4PRS* software to analyse the project schedule, construction cost, and RUC (Lee and Ibbs, 2005). This software was developed by the UC Berkeley Institute of Transportation Studies as an FHWA pooled-fund program. *CA4PRS* incorporates three interactive analytical modules: a schedule module that estimates project duration, a traffic module that quantifies the delay impact of work zone lane closures, and a cost module that compares project cost among alternatives (Lee and Choi, 2006). The results (outputs) of these three modules in *CA4PRS* integrate directly into the formulation (inputs) of LCCA.

These capabilities were confirmed on several large highway rehabilitation projects in US states including California, Minnesota, Utah, and Washington. For example, *CA4PRS* played a crucial role in the concrete pavement reconstruction of Interstate 15 Devore near San Bernardino (California), helping reduce agency cost by $8 million and saving $2 million in road user delays using continuous closures and 24/7 construction, compared with repeated (about 10 months) nighttime traffic closures, the traditional approach (Lee and Thomas, 2007).

There is growing recognition of the capabilities of *CA4PRS* and the benefits of its use. For example, *CA4PRS* won a 2007 Global Road Achievement Award from the International Road Federation (IRF). The FHWA recently endorsed *CA4PRS* as a "2008 Priority, Market-Ready Technologies and Innovations" product, and acquired an unlimited *CA4PRS* group license for all 50 states to deploy the software nationally. The American Association of State Highway and Transportation Officials (AASHTO) Technology Implementation Group (TIG) is focusing on *CA4PRS* for nationwide promotion to its members.

4.4.2 *RealCost* for LCCA

RealCost software is a LCCA tool developed by FHWA to calculate life-cycle values for both the agency and user costs associated with the construction and rehabilitation of highway projects (FHWA, 2002, 2004). The LCCA method in *RealCost* is computation intensive and ideally suited to a spreadsheet application. However, the current version of *RealCost* (version 2.1) does not have an analytical capacity to calculate agency costs or to estimate service lives for individual construction or rehabilitation activities, which should be input by users directly, based on an agency's practices. The software includes a function for automating FHWA's work zone user cost calculation method. This method for calculating user costs compares traffic demand to roadway capacity on an hour-by-hour basis, revealing the resulting traffic conditions. As with any economic tool, LCCA provides information critical to the overall decision-making process, but not the answer itself.

4.5 Pavement design (type) alternatives

In interviews with the authors, I-710 project team members who participated in the design and construction stages suggested the following three pavement design (type) alternatives for design validation from an LCCA perspective:

- Alternative 1 – innovative ACP rehabilitation: (a) Caltrans long-life CSOL for total 2.8 centreline-km and (b) long-life FDAC for total 1.6 centreline-km.
- Alternative 2 – standard ACP rehabilitation: (a) standard-life CSOL for total 2.8 centreline-km and (b) standard-life FDAC for total 1.6 centreline-km
- Alternative 3 – long-life PCCP reconstruction for 4.4 centreline-km

Based on the interviews and the Caltrans LCCA Procedure Manual (Caltrans, 2007c), more design details such as cross section, design life, and M&R sequence for each alternative were developed and confirmed by pavement experts in industry (especially, the Southern California chapter of the National Asphalt Pavement Association [NAPA] and academia [University of California, Berkeley and Davis]).

4.5.1 Alternative 1: innovative ACP

The innovative ACP alternative, which is the long-life AC pavement implemented on the I-710 project, is a new Caltrans AC pavement technology to enhance pavement quality and life (30-plus design years). It consists of two rehabilitation sections (2.8 km total) with CSOL of existing PCC slabs with AC and three FDAC replacement sections (1.6 km total) under highway overpasses. The designs for the innovative pavement structure were developed using mechanistic-empirical (ME) design methodologies to accommodate 200 million equivalent single axle loads (ESAL) over 30 years. The I-710 Long Beach project was the first demonstration of the innovative ACP in the Caltrans LLPRS program. The LLPRS was launched in 1998 to rebuild approximately 2,800 lane kilometres of deteriorated urban highways that had poor pavement structure condition and ride quality, and carried either a minimum ADT of 150,000 or average daily truck traffic of 15,000 (Lee *et al.*, 2005). The contract for the rehabilitation of LLPRS included performance-related materials specifications for stiffness, fatigue resistance, and rutting resistance for the AC layers, as well as for higher than normal compaction.

The long-life CSOL consists of 30 mm (1.2 in.) open-graded friction coarse (OGFC), 75 mm (2.9 in.) PBA-6a, 100 mm (3.9 in.) AR-8000, and 30 mm (1.2 in.) dense-graded asphalt concrete (DGAC) (Figure 4.3a). For the first FDAC sections, 640 mm (25.2 in.) of the existing PCC pavement was excavated and 150 mm (5.9 in.) of Aggregate Base (AB) was filled. The long-life FDAC consists of 30 mm (1.2 in.) of OGFC, 75 mm of PBA-6a, 150 mm of AR-8000, and 75 mm AR-8000 (+0.5 percent) (Figure 4.3b). The AR-8000 is an asphalt type having a viscosity of 8,000 poise (±25 percent) at 60°C after aging, and the AR-8000 (+0.5 percent) is a rich bottom, DGAC with conventional binder. The PBA-6a is a DGAC type with polymer-modified binder.

4.5.2 Alternative 2: standard ACP

The Standard ACP alternative is based on a typical AC pavement cross section for standard design life (20 years), which is the most common flexible pavement design in California highways. The Caltrans project team confirmed that if the innovative ACP was not proposed, then most likely the Standard ACP alternative would have been implemented to rehabilitate the existing PCC pavement on I-710 Long Beach.

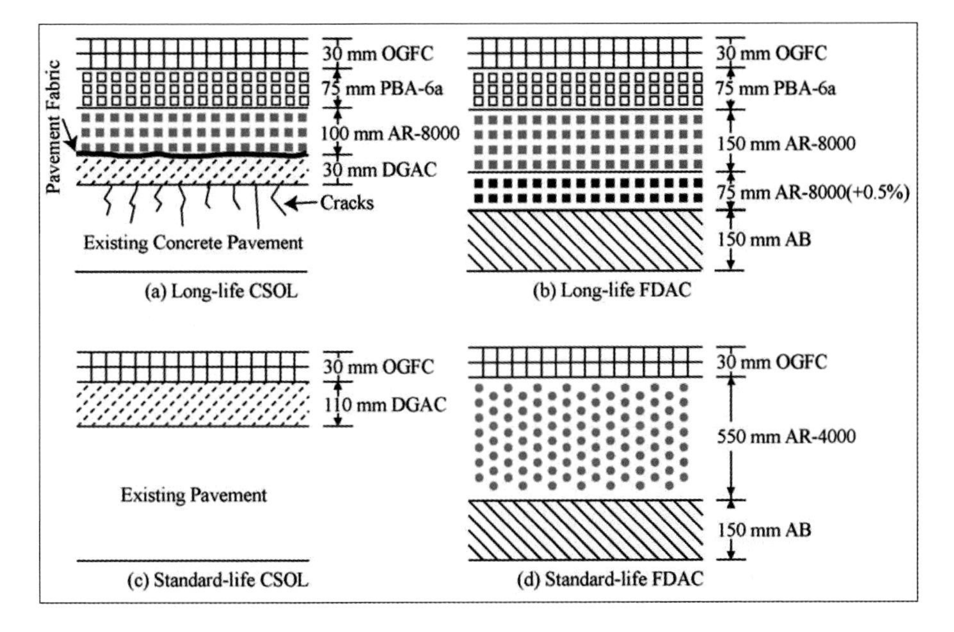

Figure 4.3 *Cross-sections of the innovative and standard ACP alternatives.*

The standard ACP alternative consisted of three standard-life FDAC replacement sections (1.6 km total) under highway overpasses and two standard-life CSOL sections (2.8 km total) between highway overpasses. For the initial construction of the Standard ACP alternative, the standard CSOL consists of 30 mm (1.2 in.) of OGFC and 110 mm (4.3 in.) of DGAC (Figure 4.3c). For the standard FDAC pavement section, 150 mm (6.0 in.) of the 730 mm (29.0 in.) of the existing pavement is excavated and filled with AB. The standard FDAC pavement consists of 30 mm (1.2 in.) of OGFC and 550 mm (22.0 in.) of Aged Residue (AR)-4000 on top of the 150 mm (6.0 in.) of new AB (Figure 4.3d). The AR-4000 is an asphalt type that has a viscosity of 4,000 poise (±25 percent) at 60°C after aging.

4.5.3 Alternative 3: long-life PCCP

Long-life PCCP competed with the innovative ACP for use in replacing the entire stretch (total 4.2 centreline-km) of the I-710 Long Beach project. The long-life PCCP consists of 300 mm (12.0 in.) of Portland concrete cement and 150 mm (6.0 in.) of hot-mix asphalt (HMA) or lean concrete base (LCB). Generally, long-life PCCP would require a slightly higher initial construction cost than long-life ACP but its use may result in lower overall maintenance and user cost over the life-cycle period because it does not demand frequent maintenance for overlays.

4.6 LCCA components

The elements required to perform an LCCA for the I-710 project (Phase 1) include pavement design alternatives, analysis period, discount rate, M&R schedules, and cost estimates. Per Caltrans policy (Caltrans, 2007c), this LCCA is based on a 60-year analysis period and is applied a fixed discount rate of 4 percent, the difference of between a 6 percent inflation rate and a 2 percent interest rate in the long-run. No salvage (or residual) value was accounted for in the 60th year and the cost of pavement treatment in the 60th year was excluded for all the alternatives. The following sections describe the major elements and their inputs and assumptions in this study.

4.6.1 Agency cost estimate

Construction cost for California highway projects consists of pavement cost, traffic-handling cost, drainage cost, specialty (storm water pollution prevention plan: SWPPP) cost, and other miscellaneous costs. In the *CA4PRS* cost module, pavement cost is estimated by a function (multiplication) of pavement item, thickness, lane width, length, and unit price. The unit prices for major pavement items were acquired from Caltrans' contractor bid database (Caltrans, 2010a). For the purpose of simplicity, other cost components, such as nonpavement items and indirect costs, and Caltrans engineering support costs were covered by using multipliers, based on Caltrans' typical cost estimate practice for rehabilitation projects, as listed in Table 4.1.

For example, the routine annualised maintenance cost used for the Standard FDAC and innovative FDAC and CSOL was $1,860 per kilometre ($3,000 per mile) and that used for the standard CSOL was $930 per kilometre ($1,500 per mile). The annual maintenance cost for the long-life PCCP was $620 per kilometre ($1,000 per mile). Routine annualised maintenance costs (dollar per lane-km per year) were acquired from the Caltrans LCCA Procedure Manual (Caltrans, 2007c).

4.6.2 Work zone user cost calculation

The weekday traffic (137,500 average daily traffic [ADT]) and weekend traffic (97,300 ADT) were collected before construction (rehabilitation) and used in the work

Table 4.1 *Typical percentages and multipliers of cost estimates.*

Number	*Construction item*	*Description*	*Percentage*	*Multiplier*
1	Pavement cost	–	100%	1.00
2	Traffic Handling cost	% of (1)	8%	1.08
3	Drainage cost	% of (1)	1%	1.09
4	Specialty (SWPPP) cost	% of (1)	15%	1.24
5	Minor cost	% of Sum (1) through (4)	5%	1.30
6	Mobilization cost	% of Sum (1) through (5)	10%	1.43
7	Supplemental cost	% of Sum (1) through (6)	5%	1.50
8	Contingency cost	% of Sum (1) through (7)	20%	1.80
9	Engineering supporting cost	% of Sum (1) through (8)	0%	1.80

Data source: Courtesy of Caltrans (2009).
Note that the construction items are specified in the *Caltrans Construction Manual*, Chapter 4. Construction details.

zone traffic analysis for the LCCA. Fifteen percent of the traffic was assumed to be truck traffic for weekdays and five percent was truck traffic for weekends. The annual growth rate of traffic volume was assumed at 0.5 percent every year (based on the California historical traffic database) (Caltrans, 2010b).

A two-peak pattern, a typical pattern for urban highway segments, was observed in both directions for weekdays. Traffic during weekends was lower than traffic on weekdays for both directions, appearing flat-peaked in the afternoon. In fact, based on this traffic pattern, Caltrans decided to undertake the project with the 55-hour extended weekend closures that least impact traffic.

In estimating RUC with the *CA4PRS* Traffic module, which is based on the *Highway Capacity Manual*'s demand-capacity model (Caltrans, 2014), traffic delay is given a road users' time value that includes the additional travel time needed to pass through a work zone and the extra time to travel through detours caused by construction activities that interfere with traffic flow. Traffic delay is converted into a dollar amount using the value of time and it is compared among the alternatives as agency cost was compared. The time value of passenger cars was $11.51 per hour and the time value of trucks was $27.83 per hour, following Caltrans policy (Caltrans, 2006).

4.7 LCCA comparison

4.7.1 Future maintenance and rehabilitation

For each pavement design alternative, future M&R details, such as sequencing, time frequency (design life), and cross sections, were developed based on the Caltrans LCCA procedure manual (Caltrans, 2007c), as summarised in Table 4.2.

4.7.2 Agency costs comparison

The agency cost of each construction activity for each alternative was also estimated using the *CA4PRS* Cost module, which incorporates the unit prices of major pavement items based on the Caltrans historical contractor bid database (Caltrans, 2010a). The agency's initial construction (rehabilitation) cost for the innovative ACP (Alternative 1) was estimated at approximately $22.6 million, and the initial construction cost of the Standard ACP (Alternative 2) was estimated at $19.3 million. The agency's initial construction cost for the long-life PCCP (Alternative 3, $43.7 million) was 190 percent higher than that of the innovative ACP.

The primary reason for the higher cost of initial construction for the long-life PCCP, compared to that of the long-life ACP, is that the concrete mix is assumed to be Rapid Strength Concrete (RSC), which cures within 12 hours from its mixing during 55-hour weekend closures. For comparison, if normal (28-day curing time mix) PCC is used, the initial construction cost might be about half of the RSC cost.

Per the LCCA procedure, future M&R construction costs were discounted (with four percent) for the NPV conversion. After applying the discount rate, the discounted total NPV of life cycle (60 years) agency cost (including the initial construction, future M&R, and annual maintenance costs) of the innovative ACP alternatives came to $33.2 million, as shown in Table 4.3 (i.e. about $18.7 million for CSOL and about $14.5 million for FDAC).

Table 4.2 *Summary of schedule estimates*

Year	*Alt. 1: innovative ACP*				*Alt. 2: standard ACP*				*Alt. 3: long-life PCCP*	
	a. CSOL		*b. FDAC*		*a. CSOL*		*b. FDAC*			
	Description	*No. of closures*	*Description*	*No. of closures*	*Description*	*No. of closures*	*Description*	*No. of closures*	*Description*	*No. of closures*
0	30 mm OGFC 75 mm PBA-6a 100 mm AR-8000 30 mm DGAC	3 weekend	30 mm OGFC 75 mm PBA-6a 150 mm AR-8000 OBC 75 mm AR-8000+0.5%	7 weekend	30 mm OGFC 75 mm DGAC 30 mm DGAC	3 weekend	30 mm OGFC 550 mm DGAC AR-4000	9 weekend	300 mm PCC 150 mm HMA or LCB	30 weekend
	1st CAPM				*1st CAPM*					
10	30 mm Mill&Rep OGFC	28 night	30 mm Mill&Rep OGFC	16 night	30 mm Mill&Rep OGFC 60 mm Mill DGAC 90 mm Rep DGAC	74 night	30 mm Mill&Rep OGFC 60 mm Mill&Fill DGAC	38 night	n.a.	n.a.
15	n.a.	n.a.	n.a.	n.a.	*2nd CAPM*					
					30 mm Mill OGFC 30 mm Mill&Rep DGAC 30 mm Rep OGFC	44 night	n.a.	n.a.	n.a.	n.a.
	2nd CAPM				*1st rehab*					
20	30 mm Mill&Rep OGFC	28 night	30 mm Mill&Rep OGFC	16 night	30 mm Mill&Rep OGFC 90 mm Mill DGAC 120 mm Rep DGAC	96 night	30 mm Mill&Rep OGFC 60 mm Mill&Fill DGAC	38 night	n.a.	n.a.

Table 4.2 *Continued*

Year	*Alt. 1: innovative ACP*				*Alt. 2: standard ACP*				*Alt. 3: long-life PCCP*	
	a. CSOL		*b. FDAC*		*a. CSOL*		*b. FDAC*			
	Description	*No. of closures*	*Description*	*No. of closures*	*Description*	*No. of closures*	*Description*	*No. of closures*	*Description*	*No. of closures*
	1st rehab				*2nd rehab*				*1st CAPM*	
30	30 mm Mill&Rep OGFC 75 mm Mill&Fill PBA-6a	76 night	30 mm Mill&Rep OGFC 75 mm Mill&Fill PBA-6a	44 night	30 mm Mill&Rep OGFC 60 mm Mill DGAC 90 mm Rep DGAC	74 night	30 mm Mill&Rep OGFC 60 mm Mill&Fill DGAC	38 night	CPR(C)	22 night
35	n.a.	n.a.	n.a.	n.a.	*3rd CAPM*				*2nd CAPM*	
					30 mm Mill OGFC 30 mm Mill&Rep DGAC 30 mm Rep OGFC	44 night	n.a.	n.a.	CPR(B)	52 night
	3rd CAPM				*3rd rehab*					
40	30 mm Mill&Rep OGFC	28 night	Mill&Rep OGFC	16 night	30 mm Mill&Rep OGFC 90 mm Mill DGAC 120 mm Rep DGAC	96 night	30 mm Mill&Rep OGFC 60 mm Mill&Fill DGAC	38 night	n.a.	n.a.

45	n.a.	n.a.	n.a.	n.a.	n.a.	n.a.	n.a.	n.a.	*3rd CAPM*	
									CPR(A)	74 night
	4th CAPM				*4th CAPM*				*1st rehab*	
50	30 mm Mill&Rep OGFC	28 night	30 mm Mill&Rep OGFC	16 night	30 mm Mill&Rep OGFC 60 mm Mill DGAC 90 mm Rep DGAC	74 night	30 mm Mill&Rep OGFC 60 mm Mill&Fill DGAC	38 night	300 mm PCC	18 weekend
55	n.a.	n.a.	n.a.	n.a.	*5th CAPM*					
					30 mm Mill OGFC 30 mm Mill&Rep DGAC 30 mm Rep OGFC	44 night	n.a.	n.a.	n.a.	n.a.

Note that weekend = 55-hour extended weekend closure (Friday 10 p.m. to Monday 5 a.m.). Night = 7-hour night time closure (9 p.m.–5 a.m.).

Table 4.3 *Summary of life-cycle costs*

Year	*Alt 1: innovative ACP*				*Alt 2: standard ACP*				*Alt 3: long-life PCCP*	
	CSOL		*FDAC*		*CSOL*		*FDAC*			
	Agency cost	*RUC*	*Agency cost*	*RUC*	*Agency cost*	*RUC*	*Agency cost*	*RUC*	*Agency cost*	*RUC*
0	$11.92M	$0.56M	$10.72M	$1.30M	$5.35M	$0.55M	$13.94M	$1.67M	$43.67M	$5.22M
	1st CAPM	1st CAPM								
10	*1st CAPM*				*1st CAPM*				n.a.	n.a.
	$1.20M	$0.58M	$0.69M	$0.33M	$3.95M	$1.57M	$1.86M	$0.80M		
15	n.a.	n.a.	n.a.	n.a.	*2nd CAPM*				n.a.	n.a.
					$1.83M	$0.83M	n.a.	n.a.		
20	*2nd CAPM*				*1st rehab*				n.a.	n.a.
	$0.81M	$0.46M	$0.46M	$0.26M	$3.51M	$1.60M	$1.26M	$0.63M		
30	*1st rehab*				*2nd rehab*				*1st CAPM (CPR C)*	
	$2.15M	$0.99M	$1.20M	$0.56M	$1.80M	$0.96M	$0.85M	$0.49M	$1.14M	$0.33M

35	n.a.	n.a.	n.a.	n.a.	*3rd CAPM*				*2nd CAPM (CPR B)*	
					$0.84M	$0.50M	n.a.	n.a.	$0.64M	$0.65M
40	*3rd CAPM*				*3rd rehab*				n.a.	n.a.
	$0.37M	$0.28M	$0.21M	$0.16M	$1.60M	$0.96M	$0.57M	$0.38M		
45	n.a.	n.a.	n.a.	n.a.	n.a.	n.a.	n.a.	n.a.	*3rd CAPM (CPR A)*	
									$1.29M	$0.82M
50	*4th CAPM*				*4th CAPM*				*1st PCC rehab*	
	$0.25M	$0.21M	$0.14M	$0.12M	$0.82M	$0.57M	$0.39M	$0.29M	$2.36M	$2.77M
55	n.a.	n.a.	n.a.	n.a.	*5th CAPM*				n.a.	n.a.
					$0.38M	$0.30M	n.a.	n.a.		
Sub-total	$16.70M	$3.08M	$13.42M	$2.73M	$20.09M	$7.84M	$18.87M	$4.26M	$49.10M	$9.79M
Annual cost	$1.97M	n.a.	$1.12M	n.a.	$0.98M	n.a.	$1.12M	n.a.	$1.29M	n.a.
Total life-cycle cost	$33.21M (agency cost) + $5.81M (RUC) =$39.02M				$41.06M (agency cost)+$12.10M (RUC) =$53.16M				$50.39M(agency cost) +$9.79M(RUC) =$60.18M	

Note that the costs in the table are discounted (with 4% rate) NPV.

The total life-cycle agency cost of the Standard ACP is about $41.1 million, whereas the long-life PCCP is about $50.4 million.

The LCCA results show that the Innovative ACP alternative requires the lowest total agency cost for the entire life-cycle analysis period (60 years) among all the alternatives. Compared with the long-life PCCP, the Innovative ACP saves a total of about $17.2 million Caltrans capital project costs (agency cost). The LCCA study justifies the implementation of the Innovative (long-life) ACP on the I-710 Long Beach project, from the total agency cost perspective in the long-run, compared with the other two alternatives.

Agency costs also include the annualised routine maintenance costs (dollar per lane-km per year), which is a relatively small dollar amount compared with the initial and major M&R costs, based on the Caltrans LCCA Procedure Manual (Caltrans, 2007c). For the annualised routine maintenance life-cycle cost of the Innovative ACP alternative, the CSOL section requires $1.97 million in total and the FDAC section requires $1.12 million in total for the 60 years of the LCCA period in NPV. For the Standard alternative, a total of $0.98 million of the annualised routine maintenance cost in NPV for the LCCA period is required for the CSOL section and $1.12 million for the FDAC section. The long-life PCCP alternative only requires $1.29 million in annualised routine maintenance costs for the 40 lane-km (25.3 lane-mile) of the PCC section.

4.7.3 Construction schedule comparison

Construction schedule (mainly the number of closures) is a required parameter for the LCCA inputs for cost estimation (especially the transportation management plan cost) and more importantly for the work zone traffic delay cost analysis.

Utilizing *CA4PRS* software, the construction schedules for initial rehabilitation and future M&R were determined for each alternative, as summarised in Table 4.2. According to the scheduling results, Alternative 1 (the Innovative ACP alternative) requires total ten (three for CSOL and seven for FDAC) 55-hour extended weekend closures and Alternative 2 (the Standard ACP alternative) requires twelve (three for CSOL and nine for FDAC) 55-hour extended weekend closures, whereas Alternative 3 (the Long-life PCCP alternative) requires thirty weekend closures for the initial construction. Comparing the schedules for the entire life-cycle analysis period (60 years), the Innovative ACP (Alternative 1) requires a total of about 188 seven-hour nighttime closures for the CSOL sections and a total of about 108 seven-hour nighttime closures for the FDAC sections for their future maintenance. The Standard ACP (Alternative 2) requires an approximate total of 546 seven-hour nighttime closures for the CSOL sections and a total of about 190 seven-hour nighttime closures of the FDAC sections for maintenance. The long-life PCCP (Alternative 3) requires another 18 weekend closures for its rehabilitation in 50 years after initial construction, in addition to a total of 148 seven-hour nighttime closures for future M&R (Table 4.2).

4.7.4 Road user costs comparison

RUC is generated by additional traffic delays due to lane closures during construction. This cost is considered as an indirect public inconvenience (time value) cost through work zone rather than an agency cost (real money) but it comes to the fore when included in an LCCA. This study utilised the *CA4PRS* traffic module. The work zone traffic analysis results show no significant difference between the user costs of the Innovative ACP (Alternative 1, $1.86 million) and the Standard ACP (Alternative 2, $2.22 million) for the initial construction. However NPV of total life cycle user cost for the Standard ACP alternative ($12.10 million) is over twice as high as that for the Innovative ACP alternative ($5.81 million) over the entire life-cycle analysis period (Table 3). The user cost of the long-life PCCP is estimated to be as much as $9.79 million, which is close to that of the Standard ACP alternative.

4.8 Summary and conclusions

LCCA for highway projects is an analytical technique that uses economic principles in order to evaluate long-term alternative investment options, especially for comparing the value of alternative pavement structures and strategies.

Life-cycle costs including agency and user costs, for three different pavement design alternatives (i.e. Innovative [long-life] ACP, Standard [-life] ACP, and Long-life PCCP) were compared with the software (*CA4PRS and RealCost)*. The LCCA utilised in the study followed the Caltrans procedure and policy, and incorporated the project team's expert opinions, collected through postconstruction interviews. Based on this information, LCCA inputs such as pavement cross sections and materials, future M&R sequencing and timeline, and lane closure schemes were generated to compare the three pavement design alternatives.

Construction schedule for initial construction and M&R activities for each of the alternatives were determined with the *CA4PRS* Schedule module. Agency costs were estimated based on material unit prices, which incorporate the Caltrans historic bid database, and pavement quantity in the *CA4PRS* cost module. User costs in the work zones for each activity were quantified in the *CA4PRS* Traffic module. The concept of NPV was used for life-cycle the cost summary and conversion with a four percent discount rate.

Agency and user life-cycle costs for the alternatives indicated that the Innovative (long-life) ACP alternative (about $39 million), which was actually implemented on the I-710 Long Beach rehabilitation project, had the lowest costs over the 60 years of the analysis period. The total life-cycle cost of the Standard ACP alternative was about $53 million and that of the Long-life PCCP Alternative was about $60 million (Figure 4.4). In summary, this LCCA case study proves that the I-710 rehabilitation project implemented the most life-cycle cost–effective pavement design type (Innovative long-life ACP), which might save a total agency and user life cycle cost of $14 million – compared to the Standard ACP alternative, which might save about $21 million, compared to the Long-life PCCP alternative – in the long run (over the 60 years of the analysis period). More specifically, the life-cycle agency cost for the Innovative ACP alternative ($33.2 million) is about $7.9 million more cost-effective than that of the Standard ACP alternative ($41.1 million), and is about $17.2 million cheaper than the Long-life PCCP alternative ($50.4 million).

This postconstruction LCCA case study not only supports and justifies the adoption of the innovative pavement technology on the I-710 Long Beach project, but it

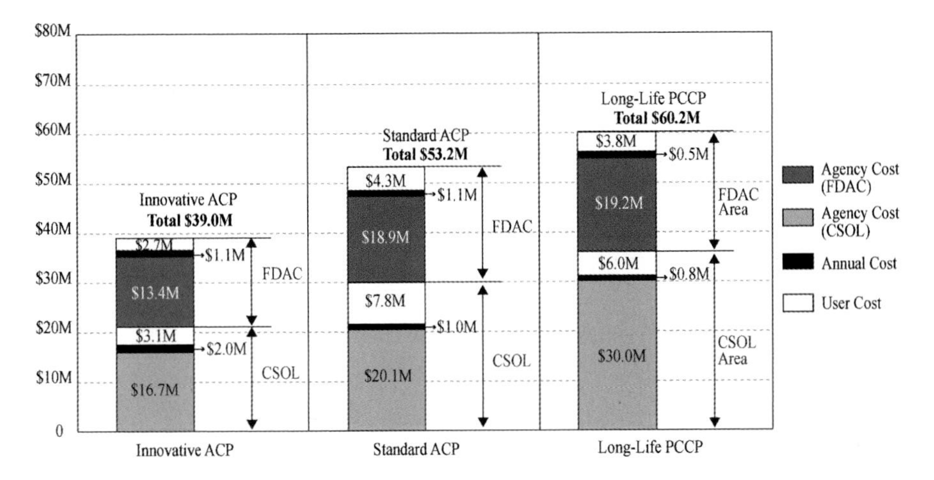

Figure 4.4 *Comparison of total life-cycle costs (NPV) for the alternatives.*

also emphasises the importance of LCCA implementation for pavement (design) type comparisons for highway rehabilitation projects.

It is recommended that transportation agencies undertake an appropriate LCCA during the pavement design and planning stages. Furthermore, utilization of construction analysis tools such as *CA4PRS* or *RealCost* will present comprehensive and realistic LCCA results, minimizing engineers' efforts and uncertainty.

Acknowledgements

This work was undertaken with funding from the California Partnered Pavement Research Program of the Caltrans, Division of Research and Innovation, which is greatly appreciated. The opinions and conclusions expressed in this chapter are those of the authors and do not necessarily represent those of California Department of Transportation or the Federal Highway Administration. The authors thank the California Department of Transportation engineers (especially, Bill Nokes) and other pavement experts (Prof. Carl Monismith and Louw Plessis) who participated in the interview and survey.

References

Caltrans (2006) *2006 travel time values for automobiles and trucks*, California Department of Transportation, Division of Traffic Operations, Memorandum. *http://www.dot.ca.gov/hq/traffops/policy/TravelTimeValues_6-30-06.pdf*. Accessed January 31, 2015.

Caltrans (2007a) *Life-cycle cost analysis*, California Department of Transportation, Office of Pavement Engineering, Sacramento, CA. *http://www.dot.ca.gov/hq/maint/Pavement/Offices/Pavement_Engineering/LCCA_index.html*. Accessed January 31, 2015.

Caltrans (2007b) *Highway design manual*, California Department of Transportation, Sacramento, CA. *http://www.dot.ca.gov/hq/oppd/hdm/hdmtoc.htm*. Accessed January 31, 2015.

Caltrans (2007c) *Life-cycle cost analysis procedures manual*, California Department of Transportation. *http://www.dot.ca.gov/hq/maint/Pavement/Offices/Pavement_Engineering/LCCA_Docs/LCCA_25CA_Manual_Final_Aug_1_2013_v2.pdf*. Accessed January 31, 2015.

Caltrans (2009) *Caltrans construction manual*, California Department of Transportation. *http://www.dot.ca.gov/hq/construc/manual2001/.* Accessed January 31, 2015.

Caltrans (2010a) *Caltrans contractor bid database,* California Department of Transportation. *http://sv08data.dot.ca.gov/contractcost.* Accessed January 31, 2015.

Caltrans (2010b) *Caltrans traffic census.* California Department of Transportation. *traffic-counts.dot.ca.gov/.* Accessed January 31, 2015.

Caltrans (2014) *Construction Analysis for Pavement Rehabilitation Strategies: Caltrans "Rapid Rehab" Software*, California Department of Transportation, Sacramento, CA. http://www.dot.ca.gov/newtech/roadway/ca4prs/. Accessed January 31, 2015.

FHWA (2002) *Life-cycle cost analysis primer.* Federal Highway Administration, U.S. Department of Transportation. *isddc.dot.gov/OLPFiles/FHWA/010621.pdf.* Accessed January 31, 2015.

FHWA (2004) *Life-cycle cost analysis RealCost user manual*. Federal Highway Administration, U.S. Department of Transportation. *http://www.fhwa.dot.gov/infrastructure/asstmgmt/rc210704.pdf.* Accessed January 31, 2015.

Gransberg, D.D. and Molenaar, K.R. (2004) Life-cycle cost award algorithms for design/build highway pavement projects. *Journal of Infrastructure Systems*, ASCE, **10**(4), 167–175.

Labi, S. and Sinha, K.C. (2005) Life-cycle evaluation of flexible pavement preventive maintenance. *Journal of Transportation Engineering*, ASCE, **131**(10), 744–751.

Lee, E.B., Lee, H. and Akbarian, M. (2005) Accelerated pavement rehabilitation and reconstruction with long-life asphalt concrete on high-traffic urban highways. *Journal of the Transportation Research Board*, No 1905, Transportation Research Board of the National Academies, Washington, D.C., 56–64.

Lee, E.B. and Ibbs, C.W. (2005) Computer simulation model: construction analysis for pavement rehabilitation strategies. *Journal of Construction Engineering and Management,* **131**(4), 449–458.

Lee, E.B. and Choi, K. (2006) Pavement rehabilitation: fast-track construction for concrete pavement rehabilitation: California urban highway network. *Journal of the Transportation Research Board*, No 1949, Transportation Research Board of the National Academies, Washington, D.C., 3–10.

Lee, E.B. and Thomas, D.K. (2007) State-of-practice technologies on accelerated urban highway rehabilitation: I-15 California experience. *Journal of Construction Engineering and Management,* **133**(2), 105–113.

Monismith, C.L., Harvey, J.T., Tsai, B., Long, F. and Signore, J. (2009) Summary Report: *The Phase 1 I-710 freeway rehabilitation project: initial design (1999) to performance after five years of traffic.* Report No: UCPRC-SR-2008-04 (FHWA No.: CA101891A), February 2009.

Papagiannakis, P. and Delwar, M. (2001) Computer model for life-cycle cost analysis of roadway pavements. *Journal of Computing in Civil Engineering,* ASCE, **15**(2), 152–156.

Salem, O., AbouRizk, S. and Ariaratnam, S. (2003) Risk-based life-cycle costing of infrastructure rehabilitation and construction alternatives. *Journal of Infrastructure Systems*, ASCE, **9**(1), 6–15.

TRB (2000) *Highway capacity manual 2000*, Transportation Research Board, National Academy of Sciences, Washington D.C.

Chapter 5
Life-cycle management framework for highway bridges

Kevin K L So and Moe M S Cheung

Summary

Transport infrastructure systems are central to economic activities and vital to economic growth. The rapid economic growth of a city leads to high demands for comprehensive and effective highway networks. The conventional procurement approach, without the integration of probabilistic life-cycle cost modelling, induces substantial long term maintenance cost. The main focus of this research is to propose a practical framework for predicting the service life by using appropriate corrosion deterioration models for reinforced concrete and fatigue models for steel bridge structures as examples. The deterioration models will then be integrated into the life-cycle cost analysis process in dealing with the long term maintenance and repair strategy. Once deterioration and life-cycle cost models of highway structures have been established, appropriate public-private partnership procurement strategies with the associated financing methods and project period determination could be developed.

5.1 Introduction to life-cycle management

5.1.1 Background

Many regions in the world are experiencing rapid growth which is symbolized by massive developments and increased construction of civil engineering infrastructure. Civil engineering infrastructure contributes to economic development both by increasing productivity and by providing amenities which enhance the quality of life. Unfortunately, infrastructure deteriorates with time and it is recognized that the expenditure for infrastructure continues well beyond its completion. In fact, statistics show that the design and construction costs constitute only a small portion of the total expenditure that society pays during the lifetime of the infrastructure. Indeed, the cost to operate and maintain infrastructure is often higher than its design and construction costs, and represents a huge investment for local and national government and taxpayers. To utilize better financial resources, it is necessary to consider this issue from a whole life cycle viewpoint. In addition, by using appropriate life-cycle management (LCM) strategies and service life prediction techniques, infrastructure performance can be improved and premature deterioration avoided, thus leading to optimum operation and maintenance of the infrastructure.

5.1.2 Overview

Transport infrastructure systems are central to economic activities and are important for economic growth. The rapid economic growth of Hong Kong, especially in last two decades, has led to very high demands for comprehensive and effective highway networks. To ensure the safety and serviceability level of the existing highway transport system, Hong Kong spends approximately HK$800 million each year in maintaining highway assets, which comprise over 1,900 km of roads, associated highway structures, roadside slopes and street furniture. According to records from the Transport Department of the HKSAR Government in December 2005, there were over 540,000 licensed vehicles and only 1,955 km of roads—436 km on Hong Kong Island, 449 km in Kowloon and 1,070 km in the New Territories. Furthermore, there are 11 major road tunnels, 1,129 flyovers and bridges, 649 footbridges and 398 subways to keep people and goods on the move. Hong Kong's roads have one of the highest vehicle densities in the world (Mak and Mo, 2005; Cheung *et al.*, 2006).

The bridge report *Bridging the Gap: Restoring and Rebuilding the Nation's Bridge* published by the American Association of State Highway and Transportation Officials (AASHTO) in 2008, identifies the serious situation the U.S. authorities are now facing. Within the next 15 years almost half of the bridges will be older than 50 years thus exceeding the lifespan for which they were designed. One in five are over 50 years old and one in four of the bridges are rated as deficient, either in need of repair or in need of widening to handle existing traffic. Even worse, new estimates in 2008 show that more than $140 billion will be needed to repair them, and soaring construction costs have whittled away the ability of states to address preventive maintenance and new bridge construction. The report summarises five main problems facing the nation's 590,000 bridges: age, congestion, soaring construction costs, lack of funds for maintenance and 'the staggering costs of new bridges'.

In a well-developed Asian country, such as Japan, there is highly developed civil infrastructure in both qualitative and quantitative terms. This includes urban expressway networks and other facilities such as bridges, dams, tunnels and so on. Because there are a huge number of civil infrastructure systems, it becomes a major social concern to develop an integrated lifetime management system for such infrastructure in the near future (Miyamoto and Frangopol, 2002). It will therefore be necessary, as in Europe and the U.S. to maintain the service life with acceptable performance and structural functions of the infrastructure stock for as long as possible and to take appropriate measures against aging effects that are in harmony with the natural environment.

5.1.3 Introduction

There has been a requirement to improve the function and performance of existing civil infrastructure beyond a set level, and to promote its beneficial use. Thus, a management system which consists of inspection, diagnosis, evaluation, prediction and countermeasure techniques needs to be developed to promote the whole life values of the infrastructure.

LCM is the application of life-cycle thinking throughout the processes of feasibility study, planning, investigation, design, construction, operation, maintenance, disposal and rehabilitation through modern business practices with the aim of managing the total life-cycle of infrastructure and services towards a more sustainable production and consumption.

LCM is not a single tool or methodology but a flexible integrated management framework of concepts, techniques and procedures incorporating the environmental, economic, and social aspects of infrastructure, processes and organisation. It is now being gradually adapted to the specific needs and characteristics of government.

Engineers and researchers try to forecast the consequences of their decisions before they proceed although they know that the odds of a forecast being correct are quite formidable. As the overwhelming majority of people in the construction industry are dealing with someone else's money, it is customary to utilize some form of accepted forecasting method to predict a result (Dale, 1993).

5.1.4 Life-cycle management of infrastructure

Conventionally, the process of infrastructure design considers each stage of construction separately and concentrates on the construction stage itself, with the primary objective of optimizing efficiency and minimizing costs during development and construction. In the past decades, much infrastructure worldwide has deteriorated and suffered from damage or defects which influence their durability. Traditional design methodology and the construction management system are the key factors that influence durability problems in which current infrastructures are unable to achieve the design life (Ma *et al.*, 2009).

For sustainable development, a traditional short-term requirement is expanded to the life-cycle of the infrastructure. As a new design methodology, infrastructure life-cycle design takes all aspects of construction practice into account, from planning, design, construction and management of service stage to demolition and the recycling of materials and components, which optimizes all alternative solutions based on the performance targets during the entire design life of the infrastructure. LCM is gaining momentum at the governmental, industry and company-wide level. This is a direct response to the changing environment and an increase in awareness and understanding fostered from LCM practices producing improved outcomes.

5.1.5 Development of life-cycle cost

Life-cycle costing (LCC) is a mathematical method used to form or support a decision and is usually employed when deliberating on a selection of options. It is an auditable financial ranking system for mutually exclusive alternatives which can be used to promote the desirable and eliminate the undesirable in a financial environment. LCC is as an assessment of all costs associated with the life-cycle of a structure or a network of structures that are directly covered by any one or more of the factors involved in the life-cycle, with complimentary inclusion of externalities that are anticipated to be internalized in the decision relevant future.

Consideration of LCC is essential when evaluating civil infrastructure construction and rehabilitation alternatives (Novick, 1990). Traditionally, LCCs are expressed as equivalent present worth of costs or as equivalent uniform annual costs, using compound interest formulas. A delicate situation is encountered when the total service life varies with respect to different rehabilitation frequencies. The benefit or the annual income resulting from the existence of structures cannot be quantified meaningfully nor can the salvage value of aging structures. The rehabilitation cost is a function of the natural extent of the degradation, location of the rehabilitation and the end user cost accrued during its nonuse. It is hence almost impossible to estimate accurately the cost prior to the completion of the rehabilitation construction (OCDE, 1983; Wang, 2005).

While the fundamentals of LCC analysis were articulated more than 100 years ago, a systematic approach appeared very much later. Typical examples of LCC analysis applications can be found in maintenance programs on highway pavement (Salem *et al.*, 2003), office buildings (ASTM, 1990; Junnila, 2004), wastewater treatment (Rebitzer, 2003), underground facilities (Kleiner, 2001), sewers (Wirahadikusumah, 2001), bridge structures (Frangopol *et al.*, 1997; So *et al.*, 2009) and bridge rehabilitation (Johnson *et al.*, 1998; Caner *et al.*, 2008), bridge management systems (Shepard *et al.*, 1993), defence industries and health services (Bull, 1993). LCC is applied intensively on roadway pavements and bridge decks (Wang, 2005).

Decisions regarding the selection of construction technology and construction materials are no longer based solely on technical and economic aspects but are becoming increasingly influenced by LCC and environmental considerations (Boussabaine and Kirkham, 2004). LCM considerations on infrastructure projects are usually integrated into different project stages including the feasibility study and planning, investigation and design, construction, operation, maintenance and the end of the service life. The following subsections are the brief highlights of LCM considerations at different project stages.

5.1.5.1 Feasibility study and planning stage

A feasibility study needs to be authentic and thorough as it is the basis for the stakeholder making an important investment decision. Major efforts during the feasibility study and planning stage may include:

1. Scope of project
   Define the needs and the scope of the proposed project for meeting the stakeholders' strategic objective and anticipated economic development. This may involve an inventory analysis in evaluating the conditions and performance adequacies of existing projects.
2. Evaluation of options
   Evaluate all solution options including their advantages and disadvantages, including the option to repair and rehabilitate the existing project.
3. Preliminary estimate of LCC
   Estimate the LCC over the whole life span of the project. In addition to the design and construction costs, it is important to include adequate operating costs, maintenance and management costs for the life of the project.

5.1.5.2 Investigation and design stage

Investigation and design are multidisciplinary processes that generally involve architects, civil engineers and mechanical engineers, as well as environmental scientists. The broad applications of LCC may be considered as follows:

1. Durable and sustainable design
   Develop and use performance-based design and specifications to improve the quality of the design and construction to achieve a durable and sustainable design by using high-performance, high-strength materials for least-size and least-weight design strategies. It reduces the size and weight of the project and the amount of construction materials and costs.

2. Long-term cost saving design
   Design efficient electrical systems and minimize the electrical loads for lighting and other electrical equipment.

5.1.5.3 Construction stage

At the construction stage of a project, there are three broad applications of LCC that may be considered (Ashworth and Hogg, 2000), as follows:

1. Method of construction
   The method of construction that the contractor chooses to employ on the project can have a major influence on the timing of cash flow and hence the time value of such payment. An appreciation of the time value of money is one of the underlying principles of LCC; therefore, it is of paramount importance that this is taken into consideration.
2. Selection of components
   Unless the design team has specifically prescribed certain elements, the contractor determines the actual selection of many components. In many circumstances, the contractor may select a component or components that comply with the specification of the design but have different effects on LCC.
3. Monitoring the LCC assumptions
   During the course of construction and particularly on large infrastructure projects, the assumptions that were used to model LCC during the design stage should be monitored.

5.1.5.4 Operation and maintenance stage

LCC techniques are used primarily as a decision making tool during the preconstruction stage of a structure. In other words, they are used to optimize the design decisions made and to inform on the likely cost implications of these decisions over the whole life. It is apparent though that, when the operation and maintenance stage of the infrastructure begins, the information obtained from the LCC analysis is unfortunately rarely used for further decision making. In fact, it can be observed that in many cases the results of the LCC analysis are simply discarded and not referred to subsequently (Boussabine and Kirkham, 2004).

Whole life costing exercises must always include forecasts of long-term, maintenance, asset replacement, operational and financial costs, as these exercises form the basis of the design decision-making process. However, during the operation and maintenance stage of the infrastructure, the management authority or stakeholders will also want to know how the infrastructure is performing economically in comparison with the original LCC forecasts. This is important because operational costs account for the largest proportion of the whole life costs. To ensure operational adequacy and efficiency during the operation and maintenance stage, the following may need to be carried out.

1. Develop a comprehensive management system
   The system may include operation and maintenance policies and regulations, budgeting, inspection requirements and performance monitoring and evaluation system.

2. Develop and maintain a performance database
   The database may include maintenance records, condition assessment data and other items that provide appropriate and timely information for decision making.
3. Reserve/contingency funds
   Provide adequate funding and resources for maintenance, periodic inspection and evaluation and proper repair and rehabilitation to achieve maximum usage of the facility.
4. Develop a maintenance plan
   Performing periodic inspection, evaluation and proper maintenance can enhance and extend the service life of the project. Extending service life instead of removal and rebuilding not only requires less natural resources and energy but also minimizes the environmental impacts. Poor-quality and poorly maintained structures will deteriorate prematurely and frequently require costly repairs, resulting in a waste of natural resources and energy.

5.1.6 Risk and uncertainty in life-cycle cost

The term risk and uncertainty are often used interchangeably, although a distinction can be drawn by noting that the concept of risk deals with measurable probabilities while the concept of uncertainty does not. An event contains an element of risk where a probability distribution can be defined. An event is uncertain when no probabilities can be developed concerning its occurrence. Risk refers to probabilities of errors in decisions and LCC forecasts throughout the life-cycle of a project, or the probabilities of occurrence of events. The more explicitly the risk is defined, the greater the possibility for the decision maker to have confidence in using the results of the LCC analysis.

In recent years, researchers began to utilise the risk-based LCC analysis approach to establish mathematical expectations of highway project benefits. For instance, Tighe (2001) performed a probabilistic LCC analysis on pavement projects by incorporating mean, variance and probability distribution for typical construction variables, such as pavement structural thickness and costs. Reigle *et al.* (2005) incorporated risk considerations into the pavement LCC analysis model. Setunge *et al.* (2005) developed a methodology for risk-based LCC analysis of alternative rehabilitation treatments for highway bridges using Monte Carlo simulation (MCS).

5.1.7 Cost and benefit analysis

Cost and benefit analysis is widely accepted as a vital support tool for economic analysis on infrastructure projects. It is primarily used to assist stakeholders for making appropriate decisions and in helping to prioritise and compare the value of different projects, not to mention helping to quantify the effect of any changes in the business environment. Usually, a cost and benefit analysis is conducted as a net present value analysis by accumulating and discounting annual cash flows associated with a particular project. On the basis of such an approach, summary statistics such as net present value, benefit/cost ratio, pay-off periods and internal rate of return can be determined. In practice, things are not usually quite so straightforward. While costs and most types of benefits can generally be quantified after a little research, an economic breakdown of all the projected benefits can sometimes be more elusive.

5.1.8 Importance of integration in life-cycle management

Deteriorating infrastructure is generally recognized where the expenditure goes well beyond its completion. In order to improve standards of living without adversely affecting the environment, the implemented solutions must be sustainable and financially viable. Recently, there has been an increasing level of awareness in research areas independently about the significance of service life prediction models, cost models and structural performances.

The requirement for service life predictions lay predominately in the estimation of maintenance, rehabilitation and life cycle replacement times. It is here that service life prediction models can provide information, either stochastically or deterministically, about likely deterioration and failure of components.

The future cost data projections are also important in the LLC comparison of different management strategies, and the structural performance condition rating recognises relationships between the latent natures of the performance of structural elements and explicitly links the deterioration of the elements to relevant explanatory variables.

However, thus far, there has been little focused effort to bring the three together in theory and practice for highway infrastructure management and the necessity for integrated LCM has been largely ignored. This chapter focuses on the missing link for highway infrastructure management in which the practical LCM strategies are integrated with three important parts: service life prediction models, cost models and structural performances.

5.2 Service life predictions

5.2.1 Introduction

Moving into an era of sustainable construction development, nowadays engineers emphasise durability-based service life design, LCM and LCC prediction. Particularly in recent decades, the long-life active use of transport infrastructure has further become a worldwide issue. The developmental spending or investment for the preparation of new infrastructure is becoming very difficult worldwide due particularly to poor economic conditions. The life extension of the transport infrastructure not only increases economic benefit, but also alleviates the financial burden on infrastructure asset management, effectively decreasing global warming and other environmental pollution in the life cycle of the structure.

Highway bridges are essential in transport infrastructure and bridge maintenance or replacement is one of the largest expenditure items in traffic infrastructure development and maintenance. Therefore, an accurate prediction of the service life of bridge structures is needed by town planners and design engineers in terms of optimizing LCCs and in maintenance schedule planning. In this section, the service life prediction models for commonly used construction materials, reinforced concrete and structural steel, in highway bridges are discussed in detail.

5.2.2 Classification of highway bridge elements

A common classification of the elements in bridges is important for the accurate reporting of conditions during inspection or after repair or rehabilitation. Moreover, an understanding of the function of various parts of bridges and their mechanisms of load transfer will have great influence on maintenance and repair procedures. Elements of highway bridges are commonly classified as substructure, superstructure and other bridge components:

1. *Substructure* includes all structural elements which support the structural deck or superstructure. These elements generally include piles, footings, caps, pedestals, piers, columns, abutments, walls and bridge seats.
2. *Superstructure* includes beams, slabs, girders or trusses.
3. *Bridge components* comprise bearings, expansion joints, parapets, road surfacing and drainage systems.

Most substructure units are constructed in reinforced concrete. In addition to examining the substructure units for cracking, spalling and other usual forms of concrete deterioration, they should also be inspected for signs of scouring, settlement, leakage, impact and accumulation of debris.

For superstructure units, bridge decks are commonly constructed in reinforced or prestressed concrete, such that decks should be examined for the usual concrete defects, including cracking and spalling. Additionally, the decks should be examined for seepage, effect of leaking expansion joins, construction blemishes, honeycombed and porous concrete, impact damage and effects of substructure or foundation movement. Prestressed concrete decks should be examined for the usual concrete defects in the same way as the examination of reinforced concrete decks. For prestressed concrete decks, special attention has to be taken to check for any signs of cracking and spalling on the prestressed section as this may indicate problems in the prestressing strands which could have significantly more severe structural consequences.

Steel beams, another important superstructure unit, should be inspected for the usual steel defects such as corrosion, deterioration of painting or galvanizing, cracking, deformation and distortion. Corrosion is by far the most widespread form of steel deterioration and the most problematic for maintenance.

Loosening of connections is most generally associated with steel truss structures. Bolts and rivets used in connections on steel structures subject to shock or impact loading tend to loosen with time. This loosening of the connection induces slip in the joints, causing distortion of the structure creating areas of extreme stress concentration, and increasing the vulnerability of the structure to fatigue failure.

Other aspects commonly found in steel beams include fatigue, abrasion and impact. Fatigue is a situation that results when a structural member is subject to repetitive or fluctuating loads which are at or below the usual allowable design values.

5.2.3 Common causes of element deterioration

The reasons leading to deterioration of the existing highway bridge stock are more or less the same in every country. The most important reasons, concerning mostly bridges which have been in service 20–30 years or more, are listed as follows (Radomski, 2002):

1. Increase in traffic flows and weight of vehicles, especially their axle loads, compared to the period when the bridge was designed and constructed.
2. Harmful influence of environmental (especially atmospheric) pollution, on the performance of structural materials.
3. Persistent use of deicing agents in countries of cold climates.
4. Low quality structural materials and bridge equipment elements, such as expansion joints, waterproofing membranes and so on.
5. Limited maintenance program or insufficient standard of maintenance, and
6. Structural and material solutions particularly sensitive to damage produced by both traffic loads and environmental factors.

In view of the importance of concrete and steel materials in highway structures, deterioration of concrete and steel structures are further elaborated in the following subsections.

5.2.4 Deterioration of concrete bridges

Reinforced concrete is widely used in bridge construction because the combination of concrete and reinforcing steel offers significant benefits over other structural materials in terms of mechanical properties, durability and aesthetics. Despite the merits of reinforced concrete structures, there are many causes for concrete deterioration. Common among these are reinforcement corrosion due to chloride attack or carbonation, freeze-thaw cycling, alkali-silica reaction and poor quality of detailing, materials or workmanship.

Steel reinforcement corrosion is a commonly encountered cause of deterioration in many concrete structures. Concrete bridge deterioration is a major problem in the operation of highway infrastructure systems (Vu and Stewart, 2000). With respect to corrosion deterioration, chloride-induced reinforcement corrosion is recognized as the major cause of deterioration of concrete bridge structures, most of which are related to bridge deck deterioration (Cady and Weyers, 1983; Louniz *et al.*, 2001). The main cause of bridge deck deterioration is the use of deicing salts in winter maintenance operations. The salt penetrates the concrete and corrodes the reinforcing steel, eventually resulting in cracking, delamination and spalling of the concrete cover. The resulting loss of both concrete and the reinforcement section area will further cause structural distress and consequent loss of load capacity (Broomfield, 1997). Similar corrosion-induced deterioration also occurs on bridge substructures, particularly piers exposed to the marine environment.

Corrosion-induced repair and rehabilitation costs are one of the most expensive items faced by highway agencies. Because of the limited funds available for upgrading and maintaining the performance of existing bridges to acceptable levels, highway agencies, governments and researchers are trying to develop models that predict optimum strategies for the maintenance planning of existing bridges, and keeping them safe and serviceable with the lowest possible investment (Yang *et al.*, 2006).

5.2.5 Deterioration of steel bridges

In Japan from the 1950s to the 1970s, a considerable amount of infrastructure was constructed to support the spreading transportation network. Much infrastructure in Japan has become old, and the number of steel bridges that have been in active service for more than 50 years is increasing dramatically and it is expected that the number of these bridges will exceed 50,000 by 2021. For these reasons, inspection, maintenance and rehabilitation planning is very important for long-life active use of bridges and infrastructure (Obata *et al.*, 2007). In Australia, there are over 30,000 road and rail bridges. For example, the Queensland government allocated $350 million towards replacing approximately 100 old and obsolete road bridges in regional areas for the period from 2006 to 2010 (Queensland, 2005). In the U.S. more than 43% of the bridges are made of steel. Currently, there are 190,000 steel bridges (simply supported and continuous) of which over 40,000 (23%) are structurally deficient and over 35,000 (18.4%) are functionally obsolete (Czarnecki and Nowak, 2008).

The majority of problems affecting the service life of bridges are related to various factors such as fatigue-sensitive details, increased service loads, corrosion deterioration and the lack of proper maintenance (Tavakkolizadeh and Saadatmanesh, 2003). Among typical types of damage (Figure 5.1) corrosion deterioration and fatigue damage of structures, particularly to steel beam bridges, are the most common which are influenced by the environment, vehicle loadings and the stress ranges.

5.2.6 Definitions of service life and limit states

Before going into details of prediction models, the terminology of service life and limit states has to be clearly defined. Service life refers to the period of time during

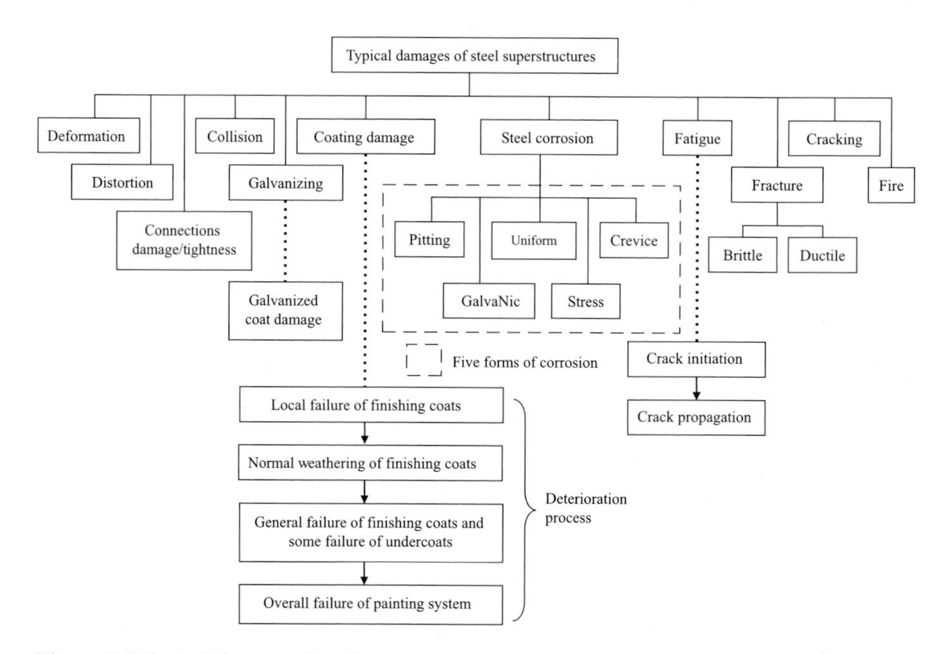

Figure 5.1 *Typical damage of steel superstructures.*

which a structure is capable of carrying out its intended function at a satisfactory performance level, with routine maintenance incorporated. As for performance, the service life of structural components can be treated differently. The necessary actions to be taken at the end of service life for bridge structural components or materials may be replacement or major repairs.

The limit states of concrete or structural steel durability usually refer to the minimum acceptable performance state or the maximum acceptable state of degradation throughout its service life. The limit states may be set with regard to ultimate limits or serviceability limit states (So *et al.*, 2009).

5.2.7 Service life prediction model for reinforced concrete structures

Service life, particularly for concrete structures, can be defined as the time at which any of the following serviceability limit states are reached: initiation of corrosion, cracking, delamination, spalling or accumulated damage reaching some specified amount. An appropriate definition of failure, and consequently service life, should consider the acceptable risk of failure, which depends on the risk involving loss of life and injury, the type of structure, the mode of failure and so forth. (Lounis and Amleh, 2004) The change of state of reinforcing steel corrosion in concrete can be referred to as the change in condition index, and expressed as a function of time.

Over recent decades, various empirical and numerical models have been proposed to estimate the deterioration caused by chloride-induced reinforcement corrosion and to predict the service life of reinforced concrete structures. One common service life model for chloride-induced corrosion of reinforcing steel suggested by Tuutti (1982) involved two stages—corrosion initiation and crack propagation. Corrosion initiation is the time when there is a sufficient concentration of chlorides at the depth of the reinforcing bar to initiate corrosion. Crack propagation is the time from corrosion damage (i.e. from initiation to cracking and spalling of the concrete cover) to the end of functional service life (Weyers *et al.*, 1993; Sarja and Vesikari, 1996; Zhang and Lounis, 2006; Chen and Mahadevan, 2008).

Once corrosion has been initiated, corrosion products build up. However, not all corrosion products contribute to the expansive pressure on the concrete, some fill the voids and pores around the steel reinforcing bar and some migrate away from the steel-concrete interface through concrete pores. Weyers (1998) concluded that the conceptual model presented by Tuutti (1982) underestimates the time to corrosion cracking when compared with the time obtained from field and laboratory observations (El Maaddawy and Soudki, 2007).

The two-stage approach for the service life of concrete was thus expanded to three stages, namely initiation, crack initiation and crack propagation. This is illustrated in Figure 5.2, which shows a modified schematic service life model illustrating different possible values of the propagation time for chloride-induced corrosion of concrete structures.

In this study, the scope of research work is limited to the most common cause of concrete deterioration—concrete structures relative to chloride-induced reinforcement corrosion.

LIVERPOOL JOHN MOORES UNIVERSITY
LEARNING SERVICES

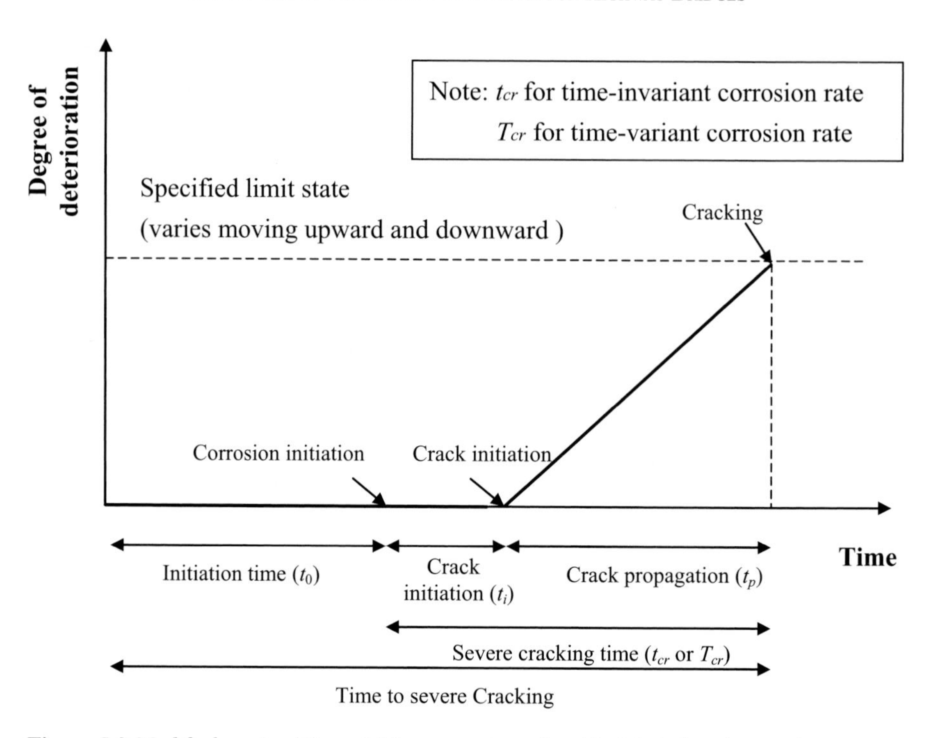

Figure 5.2 *Modified service life model for concrete under chloride-induced corrosion (Courtesy of So et al., 2009).*

The initiation time (t_0) is the time for chloride diffusion until a sufficient concentration of chlorides is available at the reinforcing bar depth to initiate corrosion. The period before the first appearance of a crack of width of ~0.05 mm is commonly called the crack initiation period (t_i). The crack propagation period (t_p) is the time from the point of crack initiation to severe cracking and spalling of the concrete cover, which basically ends the functional service life of concrete structures. The time of the three different stages can be determined by the following approach.

5.2.7.1 Determination of corrosion initiation time (t_0)

In determination of the time of corrosion initiation, several key parameters in the chloride diffusion model are identified and analysed as to their adoptability to be incorporated into the service life prediction models. These are detailed as follows:

1. *Chloride diffusion model*
   Chloride-induced corrosion is identified as the main cause of deterioration of concrete bridge structures (Lounis, 2003). The common sources of chlorides usually come from seawater or deicing salt poured on the road surface during the snow seasons, which penetrates into the concrete towards the surface of the reinforcing bars (Suwito and Xi, 2006). Chloride penetration from the environment produces a profile in the concrete characterised by high chloride content near the external surface and decreasing content at greater depths.

The chloride ingress process can be attributed to multiple transport mechanisms, the most common of which are diffusion and adsorption. However, adsorption occurs in concrete surface layers that are subject to wetting and drying cycles and only affects the exposed concrete surface down to 10–20 mm (Tuutti, 1982; Weyers *et al.*, 1993). Beyond this adsorption zone, the diffusion process dominates (*ibid.*).

The experience with both marine structures and highway structures exposed to deicing salts or seawater-spray has shown that in general this profile can be approximately modelled one-dimensionally (1-D) by Fick's second law (Duracrete, 2000; Zhang and Lounis, 2006). Field measurements by Leung and Lai (2002) have demonstrated that the adoption of the 1-D deterioration model is reasonable for service life prediction of concrete pier structures under seawater spray conditions.

Stewart and Mullard (2007) and Val and Trapper (2008) have the view that it is important to develop a two-dimensional (2-D) model for the more accurate prediction of corrosion initiation in reinforced concrete column. Cheung *et al.* (2009) have further developed a 2-D integrated corrosion performance assessment model to capture the change in environmental conditions and simulate the coupled diffusion process and the corrosion performance in the time domain, throughout the service life of the structure. Cheung's simulation approach showed practical and useful aspects in corrosion initiation time prediction, real-time bridge damage assessment and developing the life cycle maintenance plans.

However, in terms of the LCM process, it is more practical to adopt simple predictive models with the minimum number of parameters and to show a generalized situation for prediction model integration methodology. Under this consideration, a series of simulations using the real-time simulation approach suggested by Cheung *et al.* (2009) is performed and the result suggested that the long-term variation of the chloride diffusion coefficient due to change in environmental and material factors was insignificant for the purpose of this particular application and it is appropriate to adopt the averaged time-dependent diffusion coefficient approach in the prediction model.

Furthermore, the 2-D finite element simulation also indicated that, unless for very sharp corners or circular member with very small diameters, the penetration of chlorides into concrete can be reasonably modelled using the 1-D diffusion process suggested by Cheung *et al.* (2009). The proposed management strategies in the study can further be refined by integrating other sophisticated service prediction models such as Stewart and Mullard (2007), Val and Trapper (2008) and Cheung *et al.* (2009).

2. *Chloride diffusion coefficient*

   The chloride diffusion coefficient is not constant over time. Instead, it is usually expressed by means of multifactor laws by considering material properties, age of concrete, concentration and environment dependence (*ibid.*). More recent studies of existing concrete structures have shown that the decrease occurs over a long period of time and one of the main causes for this is the interaction between the concrete constituents and the various ions from the environment. The penetration and the leaching of ionic species in concrete can be studied by using a model based on the Nernst–Planck equations and a finite element procedure can then be adopted to solve the coupled nonlinear governing equation (Johannesson *et al.*, 2007; Cheung *et al.*, 2009).

As discussed in the previous subsection, this research adopts simple empirical models, but they were verified using a number of precise models. Among these models, that developed by Papadakis *et al.* (1996) is usually used. This model requires the water to cement ratio as an input parameter, which is estimated from concrete compressive strength using Bolomey's formula (Suo and Stewart, 2009). Frier and Sorensen (2005) indicated that the decrease occurs mainly during the first year of exposure, so that the coefficient can be considered constant for a mature structure. In practice, the averaged time-dependent diffusion coefficient is usually used for calculation of the induced chloride content in a given period. For simplicity, a mature concrete structure with a constant diffusion coefficient is assumed in this study.

3. *Time to corrode*

The experience with both marine structures and highway structures exposed to deicing salts or seawater-spray has shown that in general, these profiles can be approximately modelled 1-D by Fick's second law, and this approach is adopted by Duracrete (2000).

$$\frac{\partial C}{\partial t} = D\frac{\partial^2 C}{\partial x^2} \tag{5.1}$$

where C is the concentration of the diffusing substance at a distance x from the surface at time t, and D is the diffusion coefficient of the process. It is a nonstationary diffusion process. The equation is usually integrated under the assumptions that:

- the concentration of the diffusing ions, measured on the surface of the concrete, is a constant in time and is equal to C_0 ($C = C_0$ for $x = 0$ and for any y);
- the coefficient of diffusion D does not vary with time;
- the concrete is homogeneous so D does not vary with the change of the thickness of the concrete;
- the concrete does not initially contain chlorides ($C = 0$ for $x > 0$ and $t = 0$). The solution (Kirpatrick *et al.* 2002; Crank 1975; Bentur *et al.* 1997) thus obtained is:

$$C(x,t) = C_o\left[1 - erf\left(\frac{x}{2\sqrt{tD}}\right)\right] \tag{5.2}$$

The diffusion coefficient and the surface chloride content are calculated by inputting the experimental data and is often used to describe chloride profiles measured on real structures (Val and Stewart, 2003). Replacing the distance parameter x by the concrete cover thickness c, the initiation time for corrosion t_0 is obtained from the following formula:

$$C_{th} = C_0 \left[1 - erf\left(\frac{c}{2\sqrt{t_0 D}} \right) \right] \tag{5.3}$$

where C_{th} is the threshold chloride content (% by mass of cement or concrete), C_0 is surface chloride content (% by mass of cement or concrete), c is the concrete cover thickness (cm), D is the diffusion coefficient for chloride (cm²/s), t_0 is the initiation time of corrosion (s) and *erf* is the error function.

4. *Threshold chloride content*

   The threshold chloride content C_{th} is defined as the critical chloride content at which passivation of steel in concrete will be destroyed. The threshold value reportedly varies and is dependent upon many parameters such as the type of steel, the type of cement, the exposure environment (i.e. humidity, temperature and degree of oxygen saturation of the concrete) and whether the chlorides were admixed into the original concrete mix or whether they ingressed into the concrete from the external environment. The value changes depending on the weight of each parameter considered.

   For design purposes, many standards require a threshold value not higher than 0.4% of chloride by weight of cement for reinforced concrete and 0.2% for prestressed concrete. This corresponds approximately to 0.05–0.07% by weight of concrete, 0.025–0.035% for prestressed concrete.

   For example, the *European Standard BS EN 206* (2000) and the *Concrete Structure Euro-Design Handbook* (2004) restricts the chloride content to 0.2–0.4% by mass of cement for reinforced concrete, and 0.1–0.2% for prestressed concrete. *British Standard BS 8110* (1985) mentions that the maximum allowed chloride content is 0.2–0.4% chloride ions by mass of cement for reinforced concrete, and 0.1% for prestressed and heat-cured concrete.

5. *Surface chloride content*

   Field experience has shown the values of C_0 to be time dependent at the early ages but tending towards a maximum after a number of years. The chloride diffusion coefficient D for a particular concrete is considered as the rate determining parameter and is influenced by cement type, water-cement ratio, admixed or ingressed chlorides, curing, compaction, relative humidity, temperature, and time. The value of D is approximately of the order of 10^{-7}–10^{-8} cm²/s. (Sarja and Vesikari, 1996; Zhang and Lounis, 2006).

5.2.7.2 Determination of time to severe cracking ($t_0 + T_{cr}$)

Referring to Figure 5.2, the severe cracking time is the time from corrosion initiation until a specified level of corrosion induced damage is attained. These may include the serviceability limit states (e.g. excessive cracking, delamination, spalling and excessive deformation) and the ultimate limit states respectively (e.g. flexural failure, shear failure, punching shear failure) (Lounis *et al.*, 2001).

When reinforcement is placed in concrete, the alkaline condition leads to the formation of a passive layer, usually a protective thin film on the surface of the steel. However, corrosion begins if the protective thin film on the reinforcement is disrupted as a result of a falling *pH* value due to carbonation or as a result of the chloride content rising above the critical value adjacent to the embedded steel reinforcement. The products of corrosion, rust, will absorb water and occupy a volume of between two and ten times that of the steel it replaces, and thus the expansive force generated causes tensile stresses to be set up in the concrete surrounding the reinforcement. The forces generated by this expansive process can far exceed the tensile strength of the concrete, resulting in delamination, cracking and spalling of the concrete cover.

1. *Determination of crack initiation time (t_i)*
   Researchers have attempted to develop simple mathematical models to predict the time from corrosion initiation to initial cracking. Bazant (1979) introduced a mathematical model to determine the time to cracking of the concrete cover. In this model, the time to cracking is dependent mainly on corrosion rate, cover depth, spacing of steel reinforcement and the mechanical properties of the concrete, such as the tensile strength. Bazant's work was extended by Siemes *et al.* (1985) and, according to their empirical model, in the case of generalised corrosion, the critical loss of the steel reinforcement radius is based on the cracking of the cover. The cracking time suggested is dependent on three factors only—the concrete cover, the diameter of the reinforcement and the rate of corrosion. Morinaga (1988) proposed an empirical model in which the time to cracking is also a function of these three factors. It is evident that the empirical models proposed by Siemes *et al.* (1985) and Morinaga (1988) do not account for the mechanical properties of the concrete, which would significantly affect the time to corrosion cracking (El Maaddawy and Soudki, 2007). Liu and Weyers (1998) proposed a model to predict the time from corrosion initiation to the first appearance of a crack based on the amount of corrosion products required to cause cracking on the concrete cover. The main parameters used in this model are basically the same as in Bazant's model, but the concept of a porous zone around the steel reinforcing bar before creating an internal pressure on the surrounding concrete is incorporated. The period prior to the first appearance of a crack of width approximately 0.05 mm is commonly called the crack initiation period. Liu and Weyers (1998) proposed the following model to predict the time of crack initiation (t_i), which agrees reasonably well with crack initiation data (Vu *et al.*, 2005; Stewart and Mullard, 2007). Further details regarding the prediction of crack initiation time can be found in published literature (Liu and Weyers, 1998; Liang *et al.*, 2002; Branco and De Brito, 2004).

$$t_i = \frac{W_c^2}{2k_p} \tag{5.4}$$

$$k_p = 0.098\left(\frac{1}{\alpha}\right)\pi\phi i_{corr} \tag{5.5}$$

where t_i = time period for crack initiation from corrosion initiation (in years);
W_c = critical amount of corrosion products to induce cracking (mg/mm);
k_p = parameter associated with the rate of metal loss;
$Ø$ = reinforcing bar diameter (cm);

i_{corr} = corrosion rate in corrosion current density (μm/cm²); and
α = rust composition coefficient, with values between 0.5 and 0.6.

2. *Determination of severe cracking time (t_{cr} or T_{cr})*
The severe cracking time (t_{cr}) is the time from the point of corrosion initiation to a specified level of corrosion induced damage. Three main conditions occur when corrosion develops (Sarja and Vesikari, 1996), namely:

- reduction of cross sectional areas of the reinforcing steel;
- reduction of bonding between the steel reinforcement and the concrete;
- cracking of the concrete cover resulting in a reduction of the cross sectional area for the concrete load-bearing section.

Some researchers have developed mathematical models to predict the cracking time t_{cr} (Bazant 1979; Siemes *et al.*, 1985; Morinaga 1988). It is generally accepted that the service life of a structure is reduced considerably if a crack width exceeding 0.3–0.5 mm is not repaired. Results from accelerated corrosion tests of reinforced concrete slabs at the University of Newcastle confirm that the Liu and Weyers's model reasonably predicts the time for first cracking and suggests that the crack widths in the range 0.5–1.0 mm occur at approximately $10t_i$ for uniform corrosion rates (Stewart, 2001). A crack width of 0.5–1.0 mm represents severe cracking. Therefore, the cracking time is ten times the crack initiation time (Val and Stewart, 2003): $t_{cr} = 10t_i$.

The time after corrosion initiation for cracking of the concrete surface to a severe limit can be generally expressed as a time-invariant formula:

$$t_{cr} = t_i + t_p \tag{5.6}$$

$$t_{cr} \approx t_i + k_R \frac{0.0114}{i_{corr}} [A\ (c / wc)^B] \tag{5.7}$$

For $0.33\ \text{mm} \leq w_{\lim} \leq 1.0\,\text{mm}$

$$i_{corr}(t) = i_{corr}(1)\ \alpha\ (t - t_0)^{\beta}\ \ (t - t_0) \geq 1yr \tag{5.8}$$

where t_i = time period for crack initiation from corrosion initiation;
t_p = time period for crack propagation from crack initiation;
wc = water-cement ratio;
k_R = rate of loading correction factor;
i_{corr} = time-invariant corrosion current density (μA/cm²);
c = concrete cover (mm);
A, B = empirical constant with A = 70 and B = 0.23 for $w_{\lim}$ = 1.0 mm;
α, β = constants
(α = 1 and β = 0 for constant corrosion rate; and
α = 0.85 and β = –0.3 for corrosion rate reducing with time).

In fact, the corrosion rate will not be constant with time due to the formation of rust products which will reduce the diffusion of iron ions away from the steel surface (Liu and Weyers, 1998; Vu and Stewart, 2000). The time-invariant result from Equation 5.7 will then be modified with the time-variant factor incorporated. Vu *et al.* (2005) proposed the time to crack to the limit crack width (w_{lim}) for a time-variant corrosion rate as T_{cr}:

$$T_{cr} = \left[\frac{\beta+1}{\alpha}\left(t_{cr} - 1 + \frac{\alpha}{\beta+1}\right)\right]^{\frac{1}{\beta}+1} \tag{5.9}$$

for $t_{cr} > 1\,\text{yr}$ and $w_{lim} \leq 1.0\,\text{mm}$

Because Vu's model considers a time-variant corrosion rate and gives a more accurate prediction, it is adopted in this study to predict the cracking time. A limit crack width of 1.0 mm is assumed to represent severe cracking.

5.2.7.3 Determination of time for section area loss of reinforcement

After corrosion initiation, the reinforcement diameter at time t can be generalized as:

$$Ø_t = Ø_i - r_c t = Ø_i - m\lambda i_{corr}\, t \tag{5.10}$$

where $Ø_t$ = reinforcement diameter at time t (mm);
$Ø_i$ = initial reinforcement diameter (mm);
t = time after corrosion initiation (year);
r_c = rate of corrosion (mm/year) of reinforcement; and can be expressed as $m\lambda i_{corr}$,
m = corrosion coefficient dependent on type of corrosion;
λ = 0.0115 which is the factor to convert the corrosion rate from $\mu A/cm^2$ to mm/year.

A value of $m = 2$ for uniform corrosion is commonly used. For pitting corrosion, the value of m is typically 4 to 8 (Roberts and Middleton, 2000).

Stewart (2004) indicated that pitting has a more detrimental effect on the flexural strength of reinforced concrete elements. However, Melchers *et al.* (2008) pointed out that significant pitting corrosion becomes important only for highly advanced reinforcement corrosion. Li *et al.* (2008) have successfully adopted uniform corrosion-induced strength reduction in service life prediction. Recent research results (Darmawan, 2010) show that the pitting corrosion models show a less detrimental effect in flexural strength and shear strength than the general corrosion model. The above arguments regarding pitting and general corrosions are as yet inconclusive and the authors believe that pitting corrosion may have other effects on the performance of the structure, for example in ductility. However, in view of the fact that it is very difficult to predict precisely the pitting corrosion process, the general corrosion model is adopted as the base of the model in this study.

Nevertheless, in case a specific type of corrosion, such as pitting, is expected, the specific corrosion condition could be addressed in the proposed model as the amount of corrosion loss due to pitting and may significantly affect the ductility behaviour, particularly in those structures located in seismic regions. In this study, it is assumed that corrosion leads to a uniform reduction in the bar diameter of the reinforcing steel and subsequent surface damage of the concrete. As the pitting corrosion usually does not create significant spalling and induces significant intervention costs, this was not considered in this study.

For uniform corrosion, Equation 5.10 can be rewritten as:

$$t = \frac{\phi_i - \phi_t}{0.023 i_{corr}} \tag{5.11}$$

The major factors affecting the corrosion rate are the concrete quality (mainly dependent on water/cement ratio) and cover thickness. For environmental conditions with humidity of approximately 75% and a temperature of approximately 20°C, the corrosion rate is estimated in terms of a corrosion current density, i_{corr}, as follows (Vu and Stewart, 2000; Vu and Stewart, 2002; Val and Stewart, 2003):

$$i_{corr}(1) = \frac{0.378(1 - wc)^{-1.64}}{c} \times 100 \tag{5.12}$$

where wc = water/cement ratio;
c = the concrete cover thickness (mm);
$i_{corr}(1)$ = corrosion rate at the start or corrosion propagation (μA/cm²).

Most steel reinforcement corrosion reliability analyses have assumed that the corrosion rate is constant in the propagation period. However, the formation of rust products on the steel surface will reduce the diffusion of the iron ions away from the steel surface and result in the rapid decrease of the corrosion rate during the first few years after corrosion initiation, and then more slowly as it approaches a nearly uniform level (Vu and Stewart, 2000).

5.2.7.4 Validation of service life prediction models

Full-scale testing of large engineering systems for assessing performance is infeasible and expensive. With the growth of advanced computing capabilities, model-based simulation has an increasingly important role in the design of such systems. When computational prediction models are developed, the assumptions and approximations introduce errors in the code predictions.

In order to be confident of the model predictions, the computational models need to be rigorously verified and validated. When the input parameters of the model are uncertain, model prediction has uncertainty, however the validation experiments also have measurement errors.

Different prediction models for corrosion initiation, crack initiation and crack propagation are assessed critically in the study and those models adopted are validated in practice globally by various researchers. Details can be found in the literature quoted in Section 5.2.7.

5.2.8 Service life prediction models for steel structures

In conventional practice, inspection, maintenance or repair works are scheduled in the maintenance and rehabilitation manual prior to commencement of a bridge commissioning period, and the maintenance authority would conduct scheduled work, and record all the findings accordingly. However, the manual is basically a general rather than a specific manual for a particular asset. Bridges of different structural forms, at different locations or in different climates, may suffer from various degrees of deterioration. Even steel beams at different positions on the bridge, may also suffer from different degrees of damage. How to maintain effectively the bridge asset at minimum cost and how to predict the time for future work are important issues, as government funding sources become stretched.

Among typical types of steel bridge damage, especially to steel beam bridges, corrosion deterioration and fatigue damage are the most common that are influenced by the environment, vehicle loadings and the stress ranges. Corrosion and fatigue damage will weaken structural members and result in increments of deflection, reduction of ultimate bending moments, reduction in shear capacities and reduction of fatigue strength. The following models, with the integration of predefined limits, form the service life prediction models to predict the service condition of a member at any time throughout its life.

5.2.8.1 Structural steel corrosion deterioration model

Except for high performance steels such as anticorrosion weathering steel, steel beam bridges are usually subject to corrosion. If undetected over a period of time, corrosion will weaken the webs and flanges of steel beams by reducing the material thickness and possibly lead to dangerous structural failures (Lee *et al.*, 2006). In serviceability limit state analysis, measurement of the remaining thickness of the corroded steel webs and bottom flanges is commonly considered. The effective thicknesses of the webs and flanges are reduced with time as (*ibid.*):

$$t_f(t) = t_{f0} - C(t) \tag{5.13}$$

$$t_w(t) = t_{w0} - 2C(t) \tag{5.14}$$

where t_{f0} = the initial flange thickness (mm)
t_{w0} = the initial web thickness (mm)
$C(t)$ = the average corrosion penetration (mm) at time t.

Corrosion is influenced by the environment such as the amount of moisture in the atmosphere and the presence of salt. There is common agreement that the corrosion time versus penetration rate can be modelled, with a good approximation, by the exponential function (Cheung and Li, 2001):

$$C(t) = At^B \tag{5.15}$$

where $C(t)$ = average corrosion penetration in micrometres (μm) after t years;
t = time (years) of exposure;
A = corrosion loss parameter after one year of exposure;
B = parameter determined from regression analysis of experimental data.

Parameters A and B were determined by Albrecht and Naeemi (1984) and further verified by Kayser (1988), as shown in Table 5.1. Researchers have pursued extensive studies to predict time-variant corrosion propagation to capture the actual corrosion. However, these studies often neglect the influence of the periodic repainting effect on the corrosion process (Kayer, 1988; Sommer *et al.*, 1993; Jiang *et al.*, 2000; Cheung and Li, 2001; Czarnecki and Nowak, 2008).

Table 5.1 *Statistical parameters for A and B.*

Parameters	*Carbon steel*		*Weathering steel*	
	A (μm)	*B*	*A (μm)*	*B*
(a) Rural environment				
Mean value, μ	34.0	0.65	33.3	0.498
Coefficient of variation, σ/μ	0.09	0.10	0.34	0.09
Coefficient of correlation, ρ_{AB}	–	–	–0.05	–
(b) Urban environment				
Mean value, μ	80.2	0.593	50.7	0.567
Coefficient of variation, σ/μ	0.42	0.4	0.30	0.37
Coefficient of correlation, ρ_{AB}	0.68	–	0.19	–
(c) Marine environment				
Mean value, μ	70.6	0.789	40.2	0.557
Coefficient of variation, σ/μ	0.66	0.49	0.22	0.10
Coefficient of correlation, ρ_{AB}	–0.31	–	–0.45	–

Data sources: Albrecht and Naeemi (1984) and Kayser (1988).

Lee *et al.* (2006), based on previous studies, introduced a modified corrosion propagation model with periodic repainting as shown in Equation 5.16. Lee's corrosion model is adopted for service life prediction of steel beams in the study.

$$p_i(t) = C\left(t - iT_{REP} - T_{CI}\right)^m \tag{5.16}$$

$$\text{for } (i)\, T_{REP} + T_{CI} \leq t < (i+1)\, T_{REP}$$

$$p_i(t) = p_{i-1}(iT_{REP}), \text{ otherwise}$$

where $p_i(t)$ = corrosion propagation depth in micrometer (μm) at time t in years during i-th repainting period;

C = random corrosion rate parameter;

m = random time-order parameter;

T_{CI}, T_{REP} = random corrosion initiation and periodic repainting period (yrs), respectively.

5.2.8.2 Structural steel fatigue damage model

Several models have been developed to describe the process of fatigue damage, including the S-N model, Miner's linear cumulative fatigue damage model and the crack growth model under a linear-elastic fracture mechanics (LEFM) approach (Cheung and Li, 2001). S-N and Miner's models both have limitations in addressing the probabilistic nature, whereas the LEFM approach based on crack propagation theory yields more accurate results for fatigue and fracture reliability assessment if the current crack size is measured (Albrecht and Naeemi, 1984). Since the effect of crack size is taken into consideration, this approach yields more accurate results for fatigue and fracture reliability assessment if the current crack size can be measured (Kayser, 1988). In welded bridge details, the welding process inherently results in initial flaws from which crack growth can occur under cyclic loadings (Sommer *et al.*, 1993).

The commonly used crack growth model is the Paris-Erdogan model which is simplified to determine the required cycles for fatigue failure and this model is adopted in this study for service life prediction. The number of cycles required for fatigue crack growth can be estimated by integration from initial crack dimension a_0 to critical crack dimension a_c (Zhao *et al.*, 1994). Zhao's model equation is shown in Equation 5.17.

$$\int_{a_0}^{a_c} \frac{da}{[F(a,Y)\sqrt{\pi a}]^m} = \int_{N_0}^{N_c} CS^m dN \tag{5.17}$$

where $F(a,Y) = F_e F_s F_w F_g$ is defined as the geometric function of the crack; S is the stress range; and C is the crack growth constant.

and

$$\begin{aligned} \int_{N_0}^{N_c} CS^m dN &= C\sum_{i=1}^{k} S_i^{\,m}(N_{ic} - N_{i0}) \\ &= C\sum_{i=1}^{k} S_i^{\,m}(N_{ic} - N_{i0}) \\ &= CS_{rMiner}^{m} N_T \end{aligned} \tag{5.18}$$

Since randomly variable loading was involved in every case of fatigue crack propagation, an effective stress intensity range was used based on Miner's rule. The corresponding Miner's effective stress S_{rMiner} is the effective stress range while N_T is the required number of cycles to cause fatigue failure. The crack growth component m equal to 3 has been observed to be applicable to basic crack growth rate data for structural steels as well as test data on welded members. The corresponding mean value of growth constant C was found to be 1.26×10^{-13} (Zhao *et al.*, 1994; Macdougal *et al.*, 2006). Equation 5.17 can be further simplified in the study to determine the required cycles for fatigue failure:

$$N_T = \int_{a_0}^{a_c} \frac{da}{1.26\times10^{-13}[F(a,Y)S_{rMiner}\sqrt{\pi a}]^3} \tag{5.19}$$

Having estimated the fatigue life cycles, N_T, for the steel component, the data on the average daily traffic specific to the bridge superstructure, distribution of traffic volume per traffic lane and expected future traffic growth are used to convert N_T in terms of time measurement in years. If t = fatigue life in years, then,

$$N_T = (1+r)^t ADTT(365) S_c \tag{5.20}$$

where r is the yearly rate of traffic volume increase, $ADTT$ is a random variable representing the average daily truck traffic and S_c is a random variable representing an equivalent number of stress range cycles per truck crossing. The remaining fatigue life will be $t - t_0$ where t_0 is the current age of the steel component in years (Mohamadi *et al.*, 1998).

5.2.9 Risk-based service life prediction

Infrastructure performance is subject to a wide range of fluctuation because of the uncertainty existing in most of the variables such as traffic projections, material properties and climatic conditions. As a result of such uncertainties, service life is subject to random variations and should be predicted using risk analysis techniques.

Risk analysis is used to predict the probabilities of the occurrence of future events as a result of taking or not taking an action (Ang and Tang, 1975). These probabilities provide decision makers with information regarding relationships between consequences (i.e. costs, delays and safety) they might face and decisions they would make. Computer simulation can be used to derive probability distributions that describe the possible outcomes of uncertain situations (Salem *et al.*, 2003).

Traditionally, service life prediction models are fixed in nature and assume a deterministic behaviour for the service life of the infrastructure asset, since one set of variables is allowed as input and corresponding deterministic output is produced. Unfortunately, for most cases, this is not a valid assumption. The importance of including reliability concepts in the process of determining service life and subsequent process of determining associated LCC has been realized.

Service life prediction models used in this research study utilise a risk-based approach using statistical data input modelling to predict the probabilities of occurrence for alternative evaluating. The models account for uncertainty, and use input in the form of random variables sampled from the time-to-failure probability distributions developed in the probability distributions of the model outcomes, which are generated through many computer simulations. The output probability distributions can be further analysed to enable decision makers to take informed decisions regarding maintenance, rehabilitation and replacement alternatives that could be more economically applied.

5.3 Life-cycle cost models and its applications

5.3.1 Introduction

Service life prediction models would provide stakeholders with the probability of damage occurring at particular periods of time, but cannot elucidate how it would affect the structure management strategies and what actions should be taken at that predicted time to maintain the structure at a serviceable standard at minimal cost. In this study, the time-variant probabilistic LCC model is presented to predict the expected LCC of different management strategies. Results using the model can be applied to select optimal strategies for improving the durability of the structure.

5.3.2 LCC basic principle

The LCC of a structure is a combination of initial construction, inspection, maintenance, repair and failure costs occurring within the projected service life. In many cases the maintenance and repair costs can be considerably higher than the initial construction costs which include the cost of a corrosion protection system.

The net present value (NPV) method can be used to estimate the present cost of a future repair. For a repair that occurs in year t, the NPV is defined as

$$NPV = \frac{C(t)}{(1+r)^t} \tag{5.21}$$

where $C(t)$ represents the cost for a particular future action such as inspection, maintenance or repair in year t and r is the discount rate.

The minimum expected LCC has been the most widely used criterion in design optimization of new structural systems. Lifetime performance and its mathematical cost for a structure over T years of service life can be generalized as:

$$LCC = C_i + \sum_{n=1}^{T}\left[\frac{P_{ins}^n C_{ins}^n + P_m^n C_m^n + P_{rep}^n C_{rep}^n + P_f^n C_f^n}{(1+r)^n}\right] \tag{5.22}$$

where LCC = expected total life-cycle cost; C_i = initial design/construction cost and it is assumed there is no further construction during the operation and maintenance period; T = service life; P_{ins}^n = probability of inspection action conducted in year n; C_{ins}^n = expected cost of performing an inspection; P_m^n = probability of maintenance action conducted in year n; C_m^n = expected cost of routine maintenance; P_{rep}^n = probability of replacement and rehabilitation (R&R) action conducted in year n; C_{rep}^n = expected cost of R&R; P_f^n = probability of system failure in year n; C_f^t = expected cost of failure covering cost of life, user delay and so on; and r = discount rate.

5.3.2.1 Inspection costs and strategies

Inspection strategies vary depending on the policy of the stakeholders. Frequent inspections provide increased updated information for the future maintenance and repair/rehabilitation plans. However, the LCC of inspection work is usually insignificant in comparison to the whole LCC of the structure, and a cost-effective management approach should be maintained.

Inspection work (C_{ins}) may be categorized into two types: general inspection ($C_{ins,\,1}$) and detailed inspection ($C_{ins,\,2}$). General inspection is conducted annually for obvious defects which might lead to safety problems, loss of use of the structure or the restriction of use. It allows for discovering cracks, rusting or loss of material sections. Taking steel member inspection as an example, members could be visually examined regarding the condition of the protective system, including protective painting or galvanizing, the condition of materials regarding corrosion, the condition of connections, cracks, fracture defects, structural deformation or distortion.

The detailed inspection is to verify the deterioration condition of the structure and ensure it is on the schedule of the repair/rehabilitation plan. Inspection work includes visual examination of all visible and accessible parts of a structure. Some minor nondestructive testing (NDT) of representative areas, such as bridge decks and piers would take place. Thus, the detailed inspection time is adjusted, based on simulated repair/rehabilitation action times. It is usually conducted before repair/rehabilitation takes place. The LCC for inspection works can be generalized as

$$LCC_{ins} = \sum_{n=1}^{T} \frac{P_{ins}^{n} \sum_{i=1}^{L} C_{ins}^{i}}{(1+r)^{n}} \tag{5.23}$$

where L = total number of types of inspection action, and C_{ins}^{i} = cost of inspection type i.

5.3.2.2 Maintenance costs and strategies

Maintenance work can be classified primarily into two categories, preventive and corrective (Hong and Hastak, 2007). In this study, the cost associated with maintenance work (C_m) refers to routine maintenance only. Routine maintenance is conducted on a regular period covering cleansing, repainting and drainage clearing work.

Maintenance of protective paint coatings is particularly important for visual and physical preservation of steel components. A durable paint combination with long maintenance intervals is not only economical but is an environmentally acceptable solution (Saloranta, 1993). Typically, there are three types of coatings for corrosion rate reduction—paint, steel galvanizing and oxidized steel formed on weathering steel or a combination of these coatings. Because of the difficulty in galvanizing large sections of steel elements, the industry usually used paints for protection. Paints or high-performance coatings, as some of the newer systems are known, are separated into three categories: the inhibitive primer, the sacrificial primer and the barrier coat (Tam and Stiemer, 1996). Once defects on a coating surface have been found, proper cleaning of the surface and subsequent repainting should take place. The LCC for maintenance work can be generalized as

$$LCC_{m} = \sum_{n=1}^{T} \frac{P_{m}^{n} \sum_{j=1}^{M} C_{m}^{j}}{(1+r)^{n}} \tag{5.24}$$

where M = total number of types of maintenance action, and C_{m}^{j} = cost of maintenance type j.

5.3.2.3 Replacement and rehabilitation (R&R) costs and strategies

The replacement and rehabilitation work (C_{rep}) may be classified into five different actions: doing nothing ($C_{rep,1}$), minor repair ($C_{rep,2}$), major repair ($C_{rep,3}$), rehabilitation ($C_{rep,4}$) and replacement ($C_{rep,5}$).

1. 'Doing nothing' means no action is carried out and there is no change in the condition of the structure.
2. 'Minor repair' provides no improvement in durability performance but slows the deterioration rate such that the condition of the structure or its components could be maintained for a certain further period.
3. 'Major repair' provides no improvement in durability performance but restores the durability, structural strength, function or appearance of the structure. The condition of the structure or its components is improved. However, similar damage may reoccur during the remaining life of the structure such that

subsequent repair work is expected. The condition of the structure will be reset to the initial condition after major repair work.

4. 'Rehabilitation' restores the durability, structural strength, function or appearance of the structure and improves the durability performance of the repaired structure or its components. Similar damage may occur later with lower probability. The condition of the structure will be reset to the initial condition after the work.
5. 'Replacement' refers to the replacement of existing members, which may improve local capacity, durability of the structure and so forth. The condition of the structure will be reset to the initial condition after the work.

Before carrying out any rehabilitation works, the cause and severity of defects and how they affect the main structural system should be identified and analysed. Depending on the severity and type of damage caused, different rehabilitation techniques will be applied. A schematic illustration of service life extension after the proposed minor, major and rehabilitation works is shown in Figure 5.3. Possible repair and rehabilitation works are detailed with different repair codes for reinforced concrete structures, as shown in Table 5.2, proposed by So *et al.* (2009).

The repair process for structural steel elements can be complex and require the use of advanced material solutions and techniques. The major damage to the steel elements can be classified into three groups: a) corrosion destruction of members and joints, b) fatigue damage and brittle fracture, and c) mechanical fracture of the elements or the joints. Typical repair and rehabilitation methods for steel beams are proposed and illustrated in Figure 5.4 and Table 5.3, respectively. The LCC for replacement and rehabilitation work can be calculated as

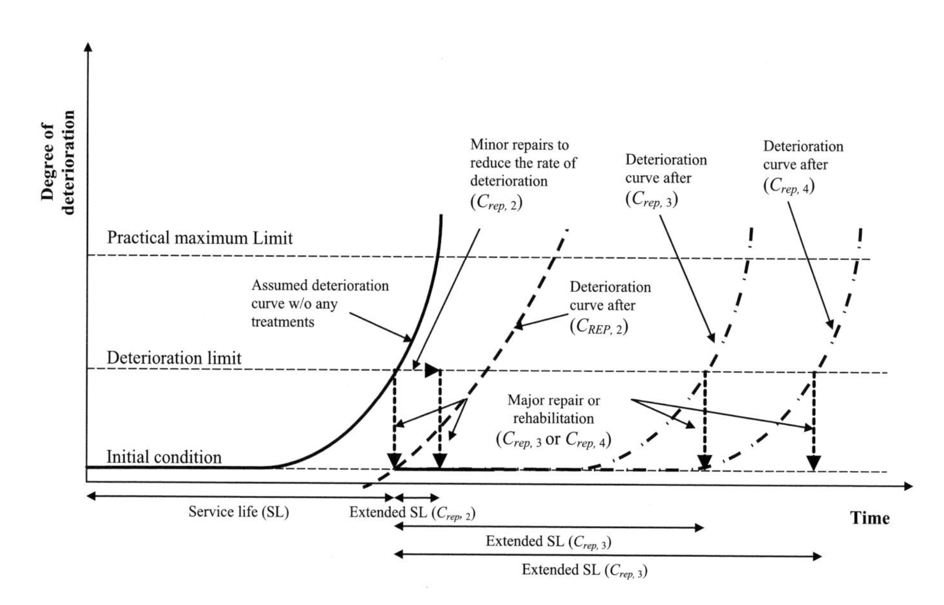

Figure 5.3 *Schematic illustration of structure condition after minor and major repair and rehabilitation works.*

Table 5.2 *Proposed repair/rehabilitation actions of defective concrete due to chloride-induced corrosion.*

Action description	*Repair code*	*Specific repair code description*	*Suggested methods*
Do nothing	0	Observation	Observation
Minor repair	1	Reduce corrosion rate	Desalination
			Coating and sealants (acrylics, rubber, copolymers, epoxy resin and polyurethanes)
	2	Repair visual defects (cracks)	Cleaning by vacuum suction and sealed by injection with the use of pressure grouting method
			Cleaning by water jetting and filled up by polymer modified cementitious mortars
Major repair	3	Removal of delamination and spalling parts	Defective concrete removal, reinforcement cleaning and patching
Rehabilitation*	4	Rebar coating	Epoxy coating or anticorrosion paints
	5	Durable repair materials	Durable material patching such as epoxy
			Apply concrete/latex modified concrete/ polyester concrete overlay
			Overall improvement by increasing concrete cover and grade
	6	Anti-corrosion surface protection	Surface painting or sheeting
			Apply hydrophobic pore liners (silicones, siloxane and silanes)
			Apply pore blockers (silicates and silicofluorides)
			Apply rendering (plain or polymer-modified cement mortars)
			Complete encapsulation
	7	System preventive measures	Installation of cathodic protection system
Replacement	8	Replace affected elements	Closure of structure and replacement by cast-in-place or new precast elements

Data source: So *et al.* (2009).
Note that rehabilitation work is to improve the durability performance such that repair methods (Codes 4–8) are usually carried out in combination.

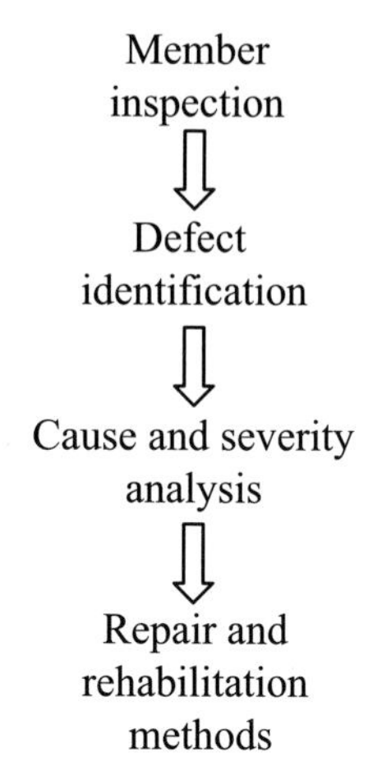

Figure 5.4 *Conventional steel girder replacement and rehabilitation works.*

Table 5.3 *Proposed repair/rehabilitation actions of defective structural steel.*

Action description	*Repair code*	*Specific repair code description*	*Suggested methods*
Do nothing	0	Observation	Observation
Minor repair	1	Reduce corrosion rate	Corrosion removal, surface cleaning, degreasing and repainting
	2	Repair visual defects	Rectify deformed parts
Major repair	3	Removal of defects	Removal/cutting out defects and repainting
Rehabilitation*	4	Repair with structural strengthening	Removal/cutting out defects and parts strengthening and repainting
	5		Structure strengthening by steel plate
	6		Structure strengthening by CFRP
	7		Structure strengthening by external tendons
Replacement	8	Replace affected elements	Closure of structures and replacement by new members

Note that rehabilitation work is to improve the durability performance such that repair methods (Codes 4–8) are usually carried out in combination.

$$LCC_{rep} = \sum_{n=1}^{T} \frac{P_{rep}^{n} \sum_{k=1}^{N} C_{rep}^{k}}{(1+r)^{n}} \tag{5.25}$$

where N = total number of types of replacement and rehabilitation action, and C_{rep}^{k} = cost of replacement and rehabilitation type j.

5.3.2.4 Salvage values of elements

The salvage values, S, for each element can be generalized (So *et al.*, 2009):

$$S^{E_{sup}} = \sum_{m=1}^{m} C_{rep,j,i_m} \frac{RSL_{eff}}{SL_{eff}} \frac{1}{(1+r)^T} = \sum_{m=1}^{m} C_{rep,j,i_m} \frac{t_{rep,j} - \Delta t}{t_{rep,j}(1+r)^T} \tag{5.26}$$

where $j = 1, 2, \ldots, l$ is an integer for types of replacement and rehabilitation action;
r = discount rate;
C_{rep,j,i_m} = last replacement and rehabilitation cost for type j on particular superstructure element i_m within life T yrs;
superstructure element Set: $E_{sup} = (i_1, i_2, \ldots, i_m)$
SL_{eff} = effective service life of element after last treatment
$= t_{rep,j}$
RSL_{eff} = remaining service life of element after last treatment beyond life T
$= t_{rep,j} - [T -$ last treatment year (yrth)]
$= t_{rep,j} - \Delta t$

5.3.3 Applications of LCC process for conventional design and construct projects

In the construction industry, the definition of a project has traditionally been synonymous with the actual construction work. As such, the preconstruction (i.e. planning and design) and postconstruction (i.e. maintenance, operation and management) activities have been sidelined, and often accelerated to reach the construction stage or to move on to the 'new job' respectively. This has resulted in poor identification of client requirements and delayed the exposure of potential solutions as needed by specialist subcontractors.

Any contemporary attempt to define or create a 'design and construction (D & C) process' will have to cover the whole 'life' of a project from recognition of a need to the operation of the completed facility. This LCM approach ensures that all issues are considered from both a business and a technical point of view. Furthermore, this approach recognises and emphasises the interdependency of activities throughout the duration of a project.

During the initial stage, the public authority uses historical LCC data to enable a broad analysis of alternative options to be made and the business case for the project to be proven. Meanwhile, the public authority may also develop and use performance-based design and specifications to improve the quality of the design and construction to achieve a durable, long-term cost saving and a sustainable design. In addition to the initial design and construction costs, it is important to incorporate adequate long-term operating costs, maintenance and management costs for the life of the project.

5.3.4 Applications of LCC process for PPP (public private partnerships) projects

Under a build-operate-transfer (BOT) project, a provider contracts to produce and maintain an asset over a defined concession period (typically 25 to 30 years), for a fee schedule determined prior to commencement of the project. At the end of the term, the ownership and operation of the asset reverts back to the public sector client.

The BOT contract may require the provider to deliver some or all of the operational and facility management services during the concession period. Almost always maintenance and replacement works are included.

A robust assessment of the relevant operational costs is of prime importance to the provider. The process and its various inputs and deliverables is complex, however the typical applications of LCC process may include the following:

1. During the initial business case stage, the public authority uses historic LCC data to enable a broad analysis of alternative options to be made and the business case for the project to be proven.
2. During the tender stage, bidders prepare LCC models, initially using high level historic data to support early financial modelling, then develop detailed LCC models as the design progresses. Several iterations of the models are likely, as the bidder refines and optimizes all cost inputs. Some 'smoothing' of annual costs (i.e. spreading of the cost peaks over two or more years to better reflect income streams) may be necessary.
3. Bidders are likely to carry out LCC assessments of alternative design/specification options at this stage.
4. During the tender stage, the public authority may also prepare its own LCC models based on an exemplar design (often referred to as a Public Sector Comparator), as a means of evaluating the LCC models submitted by the bidders.
5. Further refinement of LCCs may occur as part of the short-listing of bidders and a due diligence exercise is also likely to be carried out by both the client and the bidder's funders.
6. Once a preferred bidder is selected, the detailed LCC models setting out the expected expenditure throughout the concession period are incorporated into project financial models and contracts. For example, the provider may subcontract defined maintenance and replacement work to a third party, or the public sector client may specify in the contract replacement periods for key elements.
7. During construction and operation of the facility, the detailed LCC models are updated to reflect changes or variations in the project, or in the specification of components and materials.
8. The LCC models are regularly updated throughout the concession period and are used as a tool for managing maintenance and replacement work and for ensuring the necessary funding for such work is in place. The public authority may also use the models as a monitoring tool to ensure the necessary works are carried out.

5.3.5 Performance specifications

In order to transfer sufficient risk to the PPP awardees (the contractor), it is necessary to move from the traditional specification method for road and highway bridge maintenance to a performance specification. It is believed that a performance based payment system can allow greater flexibility, encourage innovation, enhance efficiency and improve work cost-effectiveness by allowing concurrent engineering of functions, use of new materials and technique (Zhang and Kumaraswamy, 2001).

Overall performance could be in the form of a network condition index which is a combination of residual life, skid resistance and the condition of the surface of the carriageways and footways as measured by visual condition inspections or detailed surveys.

While there is a need to have an overall performance specification, major and minor specifications are usually required to ensure that the contractor does not look after only one aspect of the overall specification such that, although they achieve the overall requirement, one or more of the other aspects do not fall below their required level of service. Specifications may also include requirements for low or optimum levels of LCCs. The following items are proposed and listed under the category of major specification and minor specification for reference.

1. Major specifications may include the remaining design life of structures, such as piers and bridge decks, remaining life of structural steel public lighting columns ... and so on.
2. Minor specifications may include surface irregularities, paint loss, wheel track rutting, edge deterioration, adverse camber, litter, landscape maintenance, breakdowns on the network, unlit lamps, road drainage, potholes, trips on the footways ... and so on.

In these types of highway PPP projects, it is intended that the contractor, as the asset manager, will have full responsibility for all issues related to the rehabilitation, maintenance and operation of the parts of the highway network included within the project. The performance standard of the contractor is then measured by a set of benchmarks as specified in the contract, For example in road markings, a continuous road marking line with more than a certain percentage loss of paint in any given section length would be counted as a defect. The lump sum payment entitled by the contractor would be adjusted according to the number of defects discovered by the public authority's representatives in the audit (Ng *et al.*, 2005).

The public authority will undertake regular inspections of the highway and shadow inspections, as well as random audits of the PPP contractors' records and databases. Surveys commissioned by the PPP contractor such as deflectographs, SCRIM, CVI will be subject to due diligence by an independent checker agreed to by all parties to the PPP contract.

5.3.6 Project financing methods

Selecting the most appropriate private sector entity is of paramount importance to a public-private partnership project. According to recent literature, the evaluation of PPP tenders should include the financial, technical, managerial, health and safety, and environment aspects. Of these, the financial package is the most important part that governments should consider (Zhang, 2004).

The recent growth of public-private partnerships is closely linked to the financing technique known as 'project finance', itself a relatively recent development. An understanding of project-finance techniques and their application in PPPs is necessary when considering policy related finance issues in PPPs.

Project finance is a method of raising long-term debt financing for major projects. It is a forum of 'financial engineering', based on lending against the cash flow generated by the project, and depends on a detailed evaluation of a project's construction, operating, and revenue risks, allocation of these between investors, lenders and other parties through contractual and other arrangements.

An adequate financial model is an essential tool for financial evaluation of a project. Some assumptions are made in the financial model for the project company and these can be classified into five main areas:

1. Macroeconomic assumptions: Include inflation, commodity prices, interest rates, exchange rates, economic growth and so on, which are not directly related to the project, but affect the financial results.
2. Project costs and funding structure: The preparation of a budget for the construction costs from the project company's point of view and determining how these are to be funded.
3. Operating revenues and costs: Closely related to the LCC of the structure or its elements. The more accurate the LCC, the better the estimation of the operating costs.
4. Loan drawings and debt service: During the construction period, the model may take into account the required equity to debt ratio, priority drawing between equity and debt, any limitations on the use of debts, drawdown schedule and so on.
5. Taxation and accounting: Although the decision to invest in a project should be based primarily on cash flow evaluation, the accounting results are important to the sponsoring company, who will not wish to show an accounting loss from investment in a project company affiliate.

5.3.7 Determination of concession period

Build-own-operate-transfer (BOOT) or BOT is a form of project financing, wherein a private entity receives a concession from the public authority to finance, design, construct and operate the facility stated in the concession contract. This enables the project proponent to pay back the loan (principal and interest) and recover their investment, operating and maintenance expenses with an expected level of profit through revenues from the project.

Owing to the long-term nature of the arrangement, fees are usually raised during the concession period. The rate of increase is often tied to a combination of internal and external variables, allowing the proponent to reach a satisfactory internal rate of return for their investment. Hong Kong's three harbour crossing tunnels are typical examples of BOT arrangements. Traditionally, such projects provide for the infrastructure, which should be in operational condition, to be transferred to the government at the end of the concession agreement.

The concession agreement generally specifies the payment structure, covenants restricting the conditions under which the public authority or the concessionaire may terminate the concession, and any compensation to be paid by one party to the other in the event of unilateral termination of the concession (Zhang and AbouRizk, 2006).

In practice, a long-term fixed concession period is the most common approach. Some countries include the construction phase as part of the concession period, others do not. There are also a few examples of concessions whose terms are variable depending on the date when the lenders recover their principal and interest, and equity holders earn a certain level of return.

Different projects will incur different cash flow profiles during their life cycles. BOT-type projects usually require a great amount of up-front investment in the construction of infrastructure facilities, the recovery of which is through revenues from the project over the concession period.

One important issue for a government considering using a BOT scheme to develop a particular infrastructure project is the determination of the appropriate length of the concession period. This length depends on a number of factors, such as the type of the project, the size and complexity of the construction activities, the operational life of the project facility, the capital structure of the concessionaire company and the market situation and revenue stream in the future operation.

There are many uncertainties and risks in construction and future operation, which have significant impact on the length of the concession period. Therefore, the concession period should be long enough to enable the concessionaire to achieve a reasonable return on their investment, but not so long that the concessionaire's return is excessive and the interests of the public sector are impaired (Zhang and AbouRizk, 2006).

5.4 Limit states, cost and performance mapping

5.4.1 Introduction

In order to relieve the long term financial burden on asset maintenance, a whole LCM concept has been introduced in the recent decade and is becoming more important in engineering design, construction and management. In the whole LCM concept, the prediction of structural member deterioration is essential for planning future maintenance actions.

One of the key success factors of LCM is how effective the service life prediction of a structure is integrated into the long-term operation and maintenance stage. The prediction of both initial service life and the effects of maintenance and rehabilitation options presents a considerable challenge. The decision-process must consider the impact of any maintenance, repair, rehabilitation or replacement options upon the future performance of the structure. However, it is easier said than done because the decision-making process is complex and indeterminate. A wrong decision may result in a future large financial burden to the local authority.

This section presents LCM strategies for assisting stakeholders to choose the best and most appropriate preventive measures for long-term maintenance or treatment for reinforced concrete bridges affected by chloride-induced corrosion deterioration and steel beam bridges affected by fatigue damage and corrosion deterioration. As such, the integration of deterioration models, corrosion deterioration and fatigue damage in the whole LCM concept are the focus. A closer analysis of illustrative LCM examples of corrosion and fatigue induced rehabilitation works is addressed. The decision analysis is referred to in the whole LCC analysis by considering appropriate elements of bridge rehabilitation costs.

5.4.2 Previous models on life-cycle management

In the last 20 years and especially in the last 10, many researchers, scientists and engineers have addressed the importance of LCM to infrastructure systems and how the service life prediction approach is integrated.

Thoft-Christensen and Sorensen (1987) proposed a reliability-based methodology to optimize inspection, maintenance and repair of infrastructure systems. Key design variables considered are inspection quality and frequency. The model minimizes the total inspection and repair costs while maintaining system reliability to an acceptable level. A service life prediction model is developed for bridge superstructures by Jiang and Sinha (1989), using the Markov chain technique and a third-order polynomial performance function. The performance function related the bridge age to an average condition rating. The model is based upon a subjective condition rating and does not permit an enhanced assessment as a result of repair and rehabilitation. Mori and Ellingwood (1992) evaluated the time-dependent reliability of reinforced concrete structures. The sensitivity of structural reliability to three degradation models is evaluated. Linear, square-root and parabolic functions are used to represent corrosion, sulphate attack and diffusion-controlled deterioration mechanisms respectively. The focus of the work is on nuclear power plants, and the authors conclude that similar approaches could be used in the evaluation of other civil structures where safety versus time is of concern. Cheung and Kyle (1996) proposed a reliability-based system for the service life prediction and management of maintenance and repair procedures of reinforced concrete structures by using statistical databases and probability theory. By using reliability as a measure of the performance requirements, inspection, preventive maintenance, repair and major rehabilitation decisions can be made based on economic analysis over the life cycle of a structure.

To achieve a comprehensive LCM strategy, Ugwa *et al.* (2005) initiated the application of an object-oriented framework to decision making in designing for durability in the bridge domain, and a prototype implementation of this framework. They discussed how an object-based solution could contribute towards achieving the objectives of durability and minimum maintenance costs at the project level.

Yang *et al.* (2006) proposed a model using the probability of satisfactory system performance during a specified time interval as a measure of reliability, treating each bridge structure as a system composed of several components in order to predict the structural performance of deteriorating structures in a probabilistic framework. Under budget constraints, it is important to prioritize the maintenance needs for bridges that are most significant to the functionality of the entire network.

Liu and Frangopol (2006) posed the network-level bridge maintenance planning problem as a combinatorial optimization. The network-level bridge maintenance planning problem is automated by a genetic algorithm to select and allocate maintenance interventions of different types among networked bridges as well as over a specified time horizon.

The LCM strategy has been successfully developed from initially dealing with major structural elements to considering the durability of whole structures, and from durability consideration of particular structures to maintenance cost optimization of systems of structures. However, there is still much work to be investigated and refined, such as the precision detection of elements' conditions, how to integrate the findings from health monitoring systems, refinement of deterioration models, incorporation of environmental costs and so on. Furthermore, one of the underlying elements in LCM, the service life prediction model, should not be overlooked. The better the service life prediction, the higher the accuracy of long-term cost estimation.

5.4.3 Dual management strategies

Conventional LCM strategy is based on fixed predefined limit states either serviceability limit state (i.e. corrosion initiation or severe cracking) or ultimate limit state (i.e. reinforcement section loss resulting in reduction of shear capacity, moment capacity, fatigue capacity, etc.). However, it may not be always true that the serviceability limit state dominates in all circumstances.

Practical LCM models based on the dual-management strategies for reinforced concretes and structural steel beams respectively are proposed. Graphical illustration of a proposed management strategy for reinforced concrete structures is shown in Figure 5.5. Also a graphical illustration of a proposed management strategy for steel beams as shown in Figure 5.6 proposes collective considerations of the serviceability limit states of deflection; the ultimate limit state of moment and shear; and the fatigue strength limit simultaneously which are essential in further LCC analysis.

As referred to previously one of the key success factors in LCM is how to predict effectively the service life and integrate it into the LCM model. In the proposed dual

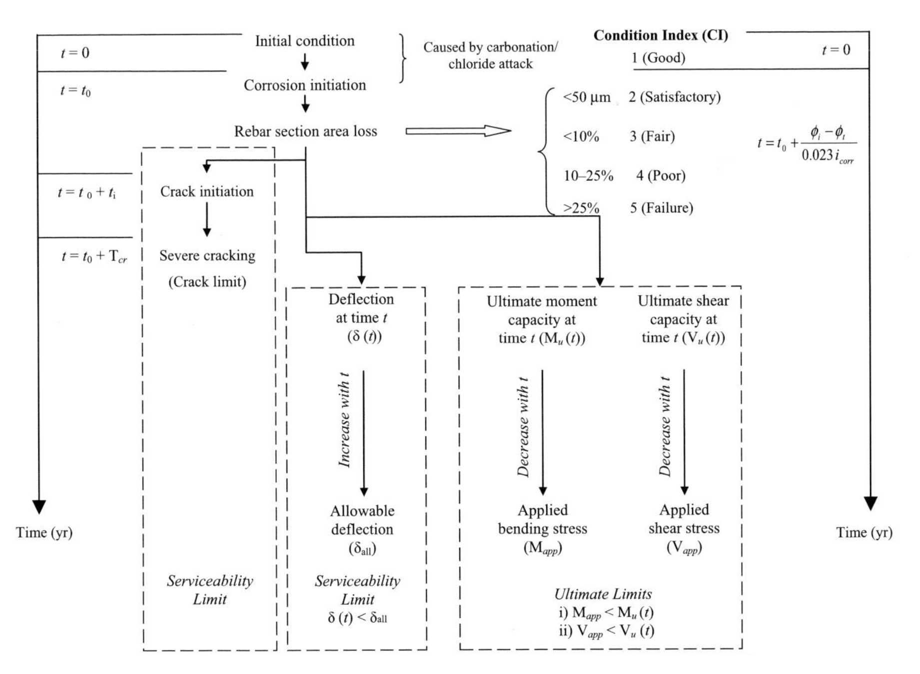

Figure 5.5 *Dual-management strategy for corrosion-induced reinforced concrete deterioration (Courtesy of So et al., 2009).*

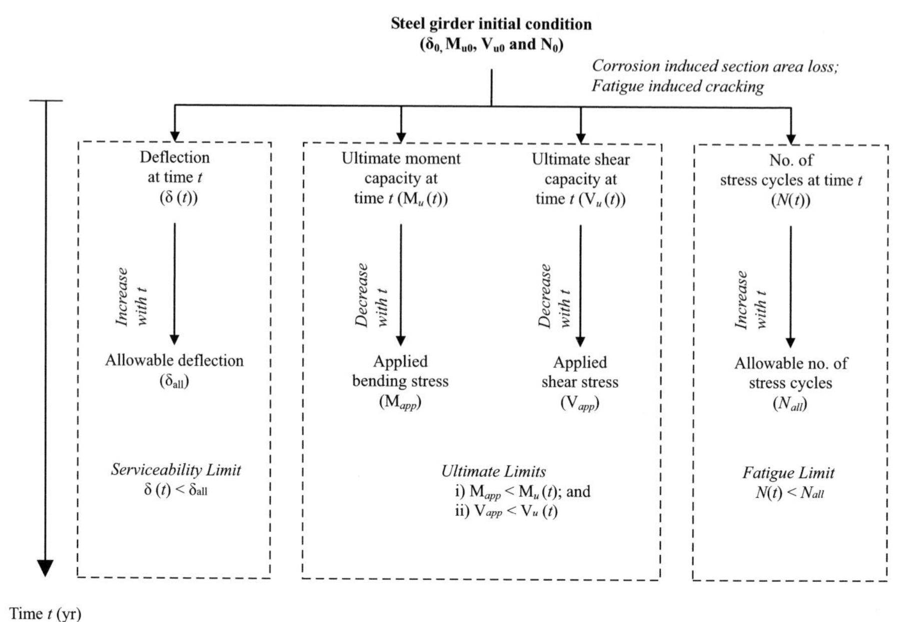

Figure 5.6 *Dual-management strategy for steel bridge girder deterioration due to corrosion and to fatigue.*

management strategies, the limit states with respect to the deterioration models will be well defined.

5.4.3.1 Reinforced concrete structures

Corrosion deterioration may result in the loss of reinforcement section areas and subsequently lead to severe cracking, increase in deflection or reduction of ultimate bending and shear capacities. Therefore, the serviceability limit states may be set according to the corrosion initiation time (t_0), crack initiation time ($t_0 + t_i$), the severe cracking time ($t_0 + T_{cr}$) and the deflection limit, while the ultimate limit state may be set according to the percentage of rebar section area loss or the extent of reduction of ultimate shear and moment capacities.

Serviceability limit states

1. *Corrosion initiation and severe cracking*

The serviceability limit states suggested in this study are defined by their visual appearance and the severity of corrosion deterioration. The time to corrosion initiation (t_0) and to severe cracking deterioration (t_0+T_{cr}) predicted by corrosion prediction models are adopted. The visual severity of corrosion deterioration is mapped with the three-stage chloride-induced corrosion model, as shown in Table 5.4. It is assumed that repair & rehabilitation works can be carried out at either the corrosion initiation time or at the severe cracking time.

Defect code no. 1 is defined as there being sufficient concentration of chlorides at the rebar depth to initiate corrosion. Defect code no. 2 is defined as any surface

Table 5.4 *Proposed degrees of deterioration for reinforced concrete structures; t_0+T_{cr} represents the time to severe cracking for a time-variant corrosion rate.*

Defect code	*Defect categories by field inspection*	*Corrosion prediction model mapping*
1	Chloride ingress of sufficient concentration	Corrosion initiation (t_0)
2	Hairline cracks < 1 mm width	Crack initiation (t_0+t_i)
3	Concrete crack > 1 mm width, map cracking, delamination and spalling	Severe cracking (t_0+T_{cr})

Data source: So *et al*. (2009).

cracks with a width less than 1 mm when inspected. Defect code no. 3 is defined as any surface cracks of a width greater than 1 mm, map cracking, delamination or cover spalling when inspected. The determination of time to corrosion, crack initiation and severe cracking are detailed in Section 2.7.

2. *Deflection*

Another serviceability limit state suggested in the study is defined in terms of the deflection of the girder at any time t in design life T. If the deflection exceeds the code requirement δ_{all}, serviceability failure is assumed to have occurred:

$$\delta_{all} \geq \delta(t) \geq \delta_0 \tag{5.27}$$

where δ_0 is the initial deflection for time $t = 0$, δ_{all} is the allowable deflection limit according to the code of practice, and $\delta(t)$ is the deflection at time t. It is assumed that the repair & rehabilitation work must be conducted before the deflection limit of the girder is reached. The deflection buffer function $F_\delta(t)$ is then formulated as

$$F_\delta(t) = \frac{\delta_{all} - \delta(t)}{\delta_{all}} \tag{5.28}$$

To assist systematic management, four proposed condition states (CS), good (G), satisfactory (S), fair (F) and poor (P) are assigned with respect to the percentage range of the deflection buffers. Let Ω_1 denote the set of possible states, then,

$$\Omega_1 = \begin{cases} G & F_\delta(t) \geq 15\% \\ S & 10\% \leq F_\delta(t) < 15\% \\ F & 5\% \leq F_\delta(t) < 10\% \\ P & F_\delta(t) < 5\% \end{cases} \tag{5.29}$$

Ultimate limit states

1. *Rebar section area loss*

The condition of the structural capacity is converted into condition states in terms of the rebar section loss due to corrosion. Let Ω denote the set of possible states for reinforced concrete structures. Then, $\Omega = [\alpha, \beta] = [0, 5]$, with 0 indicating the best

Condition Index (CI) and 5 the worst CI. Ω is a continuous set of states, which may be converted to a descriptive and discrete state set Ω′ as follows:

$$\Omega' = \{G, S, F, P, I\} \tag{5.30}$$

$$CI = \begin{cases} G & \alpha \leq CI < S_{\max} \\ S & S_{\max} \leq CI < F_{\max} \\ F & F_{\max} \leq CI < P_{\max} \\ P & P_{\max} \leq CI < I_{\max} \\ I & I_{\max} \leq CI < \beta \end{cases} \tag{5.31}$$

where G = good; S = satisfactory; F = fair; P = poor and I = failure; $S_{\max}$ = maximum numerical value of CI that belongs to category S; $F_{\max}$ = maximum numerical value of CI that belongs to category F; and $P_{\max}$ = maximum numerical value of CI that belongs to category P; $I_{\max}$ = maximum numerical value of CI that belongs to category I.

The time to rebar section loss predicted by the corrosion models is integrated into the condition states.

- G state represents the CI range from 0 to 1 referring to the time period for chloride-induced corrosion initiation.
- S state represents the CI range from 1 to 2 referring to the time period from corrosion initiation to 50 μm loss of radius.
- F state represents the CI range from 2 to 3 referring to the time period from 50μm loss of radius to a 10% area section loss.
- P state represents the CI range from 4 to 5 referring to the rebar section loss between 10% and 25%. The selection of 10% rebar loss as the lower limit of the P state is due to the change in structural behaviour being negligible when there is less than 10% section loss despite local damage of the concrete occurring (Roelfstra *et al.*, 2004).
- I state represents the CI range from 4 to 5 referring to the state of alarm, where safety is assumed to be jeopardized. Analysis of existing structures showed that a rebar section loss of 25% in crucial load carrying elements changed the structural behaviour and reduced significantly the safety margin (*ibid.*).

The proposed condition state set is used as a reference guide to benchmark the condition of the concrete structure and to prioritize the repair & rehabilitation work in the bridge networks. In the study, it is assumed that the minimum acceptable concrete condition is the F state (i.e. fair condition) and repair & rehabilitation works are conducted at the times of rebar section loss of between 10% and 25%.

2. *Ultimate shear and moment capacity*

The service limits can also be governed by the condition of the structural capacities, bending and shear, for any time t in life T:

$$M_{u0} \geq M_u(t) \geq M_{DL} + M_{SDL} + M_{LL} + M_I \tag{5.32}$$

$$V_{u0} \geq V_u(t) \geq V_{DL} + V_{SDL} + V_{LL} + V_I \tag{5.33}$$

where M_{u0} and V_{u0} are the ultimate moment and the ultimate shear capacities of the beam, respectively, at time t = 0; $M_u(t)$ and $V_u(t)$ are the ultimate moment and the ultimate shear capacities of the girder, respectively; M_{DL} and V_{DL} are moment and shear due to dead load (DL), respectively; M_{SDL} and V_{SDL} are moment and shear due to superimposed dead load (SDL), respectively; M_{LL} and V_{LL} are moment and shear due to live load (LL), respectively; and M_I and V_I are moment and shear due to impact (I), respectively. It is also assumed that repair & rehabilitation work must be conducted before the ultimate moment or shear capacity of the beam reaches the total applied moments or shears. The moment and shear buffer functions are then formulated as

$$F_M(t) = \frac{M_u(t) - M_{DL} - M_{SDL} - M_{LL} - M_I}{M_u(t)} \tag{5.34}$$

$$F_V(t) = \frac{V_u(t) - V_{DL} - V_{SDL} - V_{LL} - V_I}{V_u(t)} \tag{5.35}$$

Similarly, four proposed condition states (CS), good (G), satisfactory (S), fair (F), and poor (P) are assigned with respect to the percentage range of moment and shear buffer. Let Ω_2 and Ω_3 denote the sets of possible states for moment and shear respectively; then,

$$\Omega_2 = \begin{cases} G & F_M(t) \geq 40\% \\ S & 30\% \leq F_M(t) < 40\% \\ F & 15\% \leq F_M(t) < 30\% \\ P & F_M(t) < 15\% \end{cases} \tag{5.36}$$

$$\Omega_3 = \begin{cases} G & F_v(t) \geq 40\% \\ S & 30\% \leq F_v(t) < 40\% \\ F & 15\% \leq F_v(t) < 30\% \\ P & F_V(t) < 15\% \end{cases} \tag{5.37}$$

Options for dual management strategy

In the proposed dual management strategy, the service life limits are generalized as two options:

Option 1: serviceability limit state controls:

Corrosion initiation $$t_{\lim} = t_0 \tag{5.38}$$

Deflection $$t_{\lim} = F_\delta(t) < 5\% \tag{5.39}$$

Corrosion-induced severe cracking $$t_{\lim} = t_0 + T_{cr} \tag{5.40}$$

$$\text{if } t_0 + T_{cr} \leq t_0 + t_{10\%loss}$$

Option 2: ultimate limit state controls:

Rebar area loss $$t_{\lim} = t_0 + t_{10\%loss} \text{ if } t_0 + T_{cr} > t_0 + t_{10\%loss} \tag{5.41}$$

Shear and moment capacity at time t

$$t_{\lim} = F_V(t) \, or \, F_M(t) \leq 15\% \tag{5.42}$$

5.4.3.2 Steel structures

Corrosion deterioration and fatigue damage may weaken structural members and result in increased deflection, reduction of ultimate bending and shear strengths and reduction of fatigue strength. The corrosion deterioration and fatigue damage models, with integration of predefined ultimate limits for shear, bending and fatigue and serviceability limits for deflection, form service life prediction models to predict the service condition of steel members at any time throughout their life.

Serviceability limit states

It is defined in terms of deflection of a beam at any time, $\delta(t)$, in limit design life. If the deflection exceeds the code requirement, δ_{all}, a serviceability failure is assumed to occur. It is assumed that repair & rehabilitation work must be conducted before a beam reaches its deflection limit.

Ultimate limit states

1. *Ultimate shear and moment capacity*

The ultimate limits are governed by the structural capacity condition in bending, M, and shear, V, for any time in the life span. It is assumed that repair & rehabilitation work must be conducted before the ultimate moment or shear capacity of the beam, $M_u(t)$ and $V_u(t)$, reaches the total applied moment or shear.

2. *Ultimate fatigue strength capacity*

The fatigue strength limit suggested in the study is defined in terms of the accumulated number of stress cycles at any time, $N(t)$, throughout its life. If the number of stress cycles exceeds its allowable, N_{all}, a fatigue failure is assumed to occur. Similarly, it is assumed that repair & rehabilitation work must be conducted before reaching the limit.

Mathematical models

To present mathematically the condition of structural members in regard to their limits at a particular time, buffer functions, $F(t)$, measured in percentage for each limit are formulated according to equations 5.43 to 5.46, where subscripts δ, M, V and N for $F(t)$ refer to buffer functions for deflection, moment, shear and fatigue strength respectively. Subscripts *DL, SDL, LL* and *I* for M or V refer to moment or shear induced by dead load, superimposed dead load, live load and impact load, respectively.

$$F_\delta(t) = \frac{\delta_{all} - \delta(t)}{\delta_{all}} \tag{5.43}$$

$$F_M(t) = \frac{M_u(t) - M_{DL} - M_{SDL} - M_{LL} - M_I}{M_u(t)} \tag{5.44}$$

$$F_V(t) = \frac{V_u(t) - V_{DL} - V_{SDL} - V_{LL} - V_I}{V_u(t)} \tag{5.45}$$

$$F_N(t) = \frac{N_{all} - N(t)}{N_{all}} \tag{5.46}$$

Assignment of condition states to the steel beam

Four proposed condition states, good (G), satisfactory (S), fair (F) and poor (P) are assigned to the percentage range of the deflection, moment, shear and fatigue strength buffer, respectively. Let Ω_j denote the set of possible limit states, where j = 1, 2, 3 and 4 with respect to different limits: deflection, bending moment, shear and fatigue, respectively.

For deflection:
$$\Omega_1 = \begin{cases} G & F_\delta(t) \geq 15\% \\ S & 10\% \leq F_\delta(t) < 15\% \\ F & 5\% \leq F_\delta(t) < 10\% \\ P & F_\delta(t) < 5\% \end{cases} \tag{5.47}$$

For bending moment:
$$\Omega_2 = \begin{cases} G & F_M(t) \geq 40\% \\ S & 30\% \leq F_M(t) < 40\% \\ F & 20\% \leq F_M(t) < 30\% \\ P & F_M(t) < 20\% \end{cases} \tag{5.48}$$

For shear:
$$\Omega_3 = \begin{cases} G & F_V(t) \geq 40\% \\ S & 30\% \leq F_V(t) < 40\% \\ F & 20\% \leq F_V(t) < 30\% \\ P & F_V(t) < 20\% \end{cases} \tag{5.49}$$

For fatigue strength:
$$\Omega_4 = \begin{cases} G & F_N(t) \geq 40\% \\ S & 30\% \leq F_N(t) < 40\% \\ F & 20\% \leq F_N(t) < 30\% \\ P & F_N(t) < 20\% \end{cases} \tag{5.50}$$

The percentage ranges of buffers for different limit states are variously dependent on the stakeholders' decisions. In this study, it is assumed that the 'F' condition state is the minimum acceptable condition level. Once acceptable service condition states for each limit have been predefined, future replacement or rehabilitation works should be carried out before reaching its acceptable condition limits. The predicted action time will then be adopted in the LCC model for the subsequent cost-benefit analysis.

Options for dual management strategy

In the proposed dual management strategy, the service life limits are generalized as two options:

Option 1: serviceability limit state controls:

Deflection $$t_{\lim} = F_{\delta}(t) < 5\% \tag{5.51}$$

Option 2: ultimate limits control:

Shear and moment capacity and fatigue strength at time t

$$t_{\lim} = F_V(t), F_M(t) \quad \text{or} \quad F_N(t) \leq 20\% \tag{5.52}$$

5.5 Application of life-cycle management strategies

5.5.1 Application in infrastructure asset management

The concept of infrastructure asset management is becoming increasingly important for those responsible for managing highway networks. Asset management is not a new concept and most highway authorities are practicing elements of asset management already (Country Surveyors' Society, 2004).

The number of deteriorating highway infrastructures is increasing worldwide. Costs of inspection, maintenance, replacement and rehabilitation of these infrastructures far exceed available budgets. Maintaining the safety of existing bridges by making better use of available resources is a major concern in bridge management. Internationally, the engineering profession has taken positive steps to develop more comprehensive bridge performance measures, conduct better maintenance activities and provide new forms of bridge management. Thus, nowadays, the concept of integrated LCM and LCC analysis plays an important role in asset management.

Effective life cycle planning is about making the correct investment at the correct time to ensure that the asset delivers the requisite level of service over its full expected life, at minimum cost. Meanwhile, life cycle benefits of this investment must be maximized to ensure the needs of society are optimally met, taking into consideration safety, economy and sustainability requirements.

The plan should cover the expected life of the component from new to replacement or, for indefinite life components, the life of the treatment cycle from an 'as new' condition back to an 'as new' condition. The plan should also include the timing, nature and cost of all treatments needed to maintain the service potential of the asset, component or group of components over their useful life.

The following illustrated examples are used to demonstrate the practical application of the proposed integrated LCM strategies and how the LCM concept can be used in realistic infrastructure asset management for future maintenance treatments selection, priority listings and so on.

5.5.1.1 LCM Strategy for RC decks and piers of highway bridges Service life prediction

To demonstrate the application of the proposed LCM strategy (Figure 5.7) on corrosion-induced reinforced concrete deterioration, a reinforced concrete bridge deck and piers are used as examples. Both elements considered here are designed for a 100-year

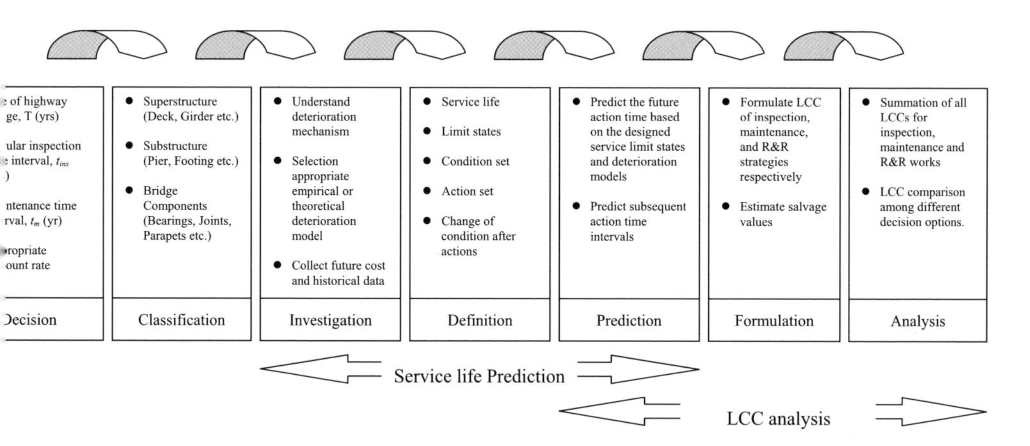

Figure 5.7 *Suggested LCM flows for inspection, maintenance and replacement & rehabilitation (R&R) works.*

life. The bridge deck and piers are under different exposure conditions, and are subject to deicing agents and seawater respectively. The statistical parameters for service life simulation are shown in Table 5.5.

Other relevant material properties include the density of the corrosion products, the density of steel, the thickness of pore zone concrete–steel, Poisson's ratio for concrete and the creep coefficient of concrete. These parameters are assumed to be 3,600 kg/m^3, 7,800 kg/m^3, 12.5μm, 0.18, 2.0 respectively. The corrosion product composition was assumed to have a value between 0.5 and 0.6.

One of the major factors affecting the rate of chloride ion penetrating into concrete is the chloride diffusion coefficient which varies in different climatic regions and with the quality of workmanship. The chloride diffusion coefficient for the bridge deck adopted in the study is obtained from Val and Trapper (2008). This is a typical value of the chloride diffusion coefficient of ordinary Portland cement with a water-to-cement (w/c) ratio of 0.5. The chloride diffusion coefficient for the bridge pier adopted in the study is obtained from field measurements in Hong Kong on pier concrete with a 0.45 w/c ratio (Leung and Lai, 2002). Based on the chloride-induced corrosion models and statistical parameters, the cumulative probabilities of different defined limit states at the times of corrosion initiation (t_0) and severe cracking (t_0+T_{cr}), and the times in terms of the rebar section area loss ($t_{10\%\ loss}$) at the bridge deck and the tidal and splash zones of the piers are simulated using the MCS method, as shown in figures 5.8–5.10.

The mean and 80 percentile simulated values are listed in Tables 5.6 and 5.7 and are employed in the subsequent LCC analysis. The impacts of rebar section area loss to deflection, shear and moment capacities are for simplicity neglected in the examples. The adopted percentile value is dependent on the stakeholders' decision. An 80 percentile value used in the study is in accordance with the Hong Kong local practice in risk simulation.

Life-cycle cost comparison

The simulation results are incorporated with the cost data in Table 5.8 to calculate the LCC of the bridge deck and the pier. For both examples, conventional carbon steel reinforced concrete of grade 45 with a 45mm cover thickness is set as the reference

Table 5.5 *Statistical parameters of random variables used for the illustrative examples.*

Parameters	*Major affecting factors*	*Mean*	*C.O.V* [1]	*Distribution*	*Sources*
Surface chloride content (C_0)	Environment	Pier—splash zone 1.44 wt% of concrete	0.7	LN[2]	Val (2005) Leung and Lai (2002)
		Pier—tidal zone 1.56 wt% of concrete			
		Deck (with deicing agent) 5 kg/m³	0.2	LN	Mangat (1991)
Diffusion coefficient (D)	Concrete quality	Pier—splash zone $D = 10.32 \times 10^{-13}$ m²/s	0.10	LN	Leung and Lai (2002) Enright and Frangopol (1998)
		Pier—tidal zone $D = 9.34 \times 10^{-13}$m²/s			
		Deck: $D = 6 \times 10^{-12}$ m²/s	0.2	LN	Val and Trapper (2008)
Threshold chloride content (C_{th})	Steel quality	Pier: 0.033 wt% of concrete	0.10	LN	Bhaskaran *et al*. (2006)
		Deck: 0.9 kg/m³	0.19		Stewart and Rosowsky (1998)
Concrete cover thickness (c)	Workmanship	Specified (45 mm or 50 mm)	0.20	LN	Sudret (2008)
Diameter of rebar (Ø)	Fabrication process	25 mm	0.02	LN	Enright and Frangopol (1998)
Concrete elastic modulus (Ec)	Concrete quality	37,222 MPa	0.12	N[3]	Li (2003); Melchers *et al*. (2008)
Concrete strength (f'_c)	Workmanship	45.8 MPa	0.135	LN	Li (2003); Biondini *et al*. (2006)
Water-to-cement ratio (wc)	Workmanship	0.45 for Pier	0.1	N	Leung and Lai (2002)
		0.50 for Deck			Val and Trapper (2008)

[1] C.O.V. – Coefficient of variation.
[2] LN – Lognormal distribution.
[3] N – Normal distribution.

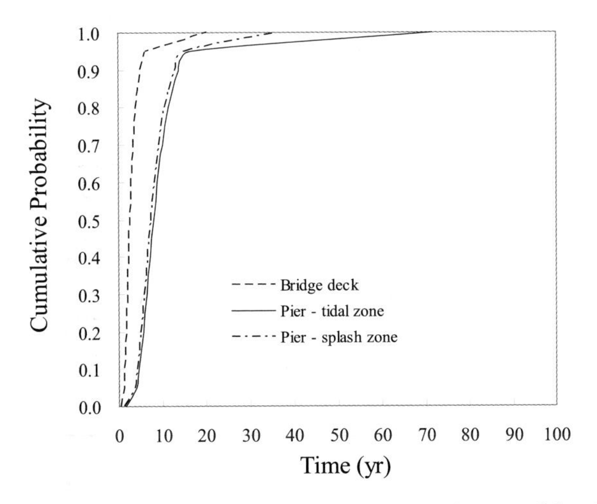

Figure 5.8 *Probability of corrosion initiation (t_0) in bridge deck and pier – tidal and splash zones.*

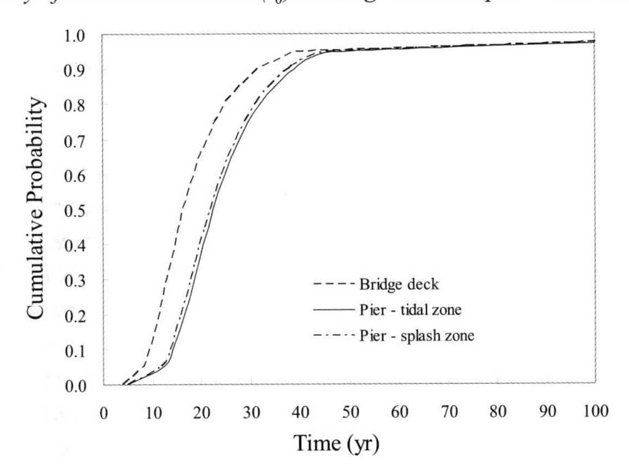

Figure 5.9 *Probability of severe cracking (t_0 + T_{cr}) in bridge deck and pier – tidal and splash zones.*

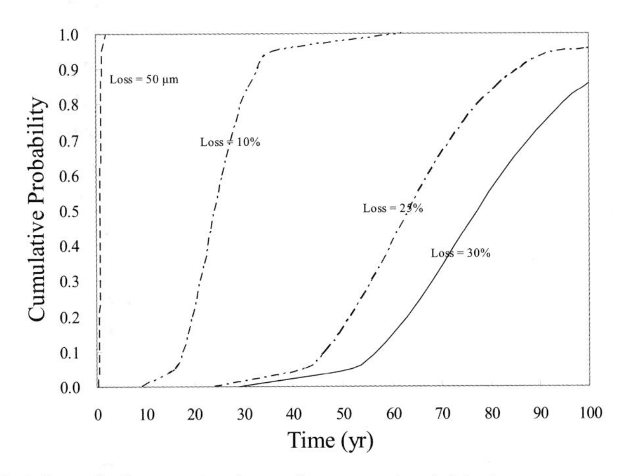

Figure 5.10 *Probability of rebar section loss after corrosion initiation.*

Table 5.6 *Summary of times to corrosion initiation and cracking (yrs).*

Serviceability limit states	*Value*	*Locations*		
		Bridge deck	*Pier—splash zone*	*Pier—tidal zone*
Corrosion initiation (t_0)	Mean	3.07	8.14	8.85
	80% tile	4.07	10.62	11.58
Severe cracking (t_0+T_{cr})	Mean	19.21	24.55	25.19
	80% tile	25.00	31.13	32.04

Table 5.7 *Summary of 25 mm diameter rebar area section loss (yrs) after corrosion initiation.*

Section loss of rebar	*Value*	*Year*
50 μm of radius	Mean	0.97
	80% tile	1.16
10% Loss	Mean	24.94
	80% tile	29.70
25% Loss	Mean	65.12
	80% tile	81.20
30% Loss	Mean	79.40
	80% tile	94.54

case. An improved durability carbon steel reinforced concrete of Grade 50 with a 70mm cover thickness, called the strengthened case, illustrates the effect of the defined service life limit state to LCC. Alternative reinforcing materials (i.e. stainless steel and epoxy-coated steel) are used to demonstrate the cost justification.

1. *LCC example 1: deicing agent attacks on bridge decks*

For many concrete bridges located in cold regions, bridge decks will need to be repaired or rehabilitated well before other elements of the bridge structures. Application of deicing salts resulting in chloride-induced rebar corrosion is commonly known as a major cause of premature deterioration of bridge decks. The results of the mean LCC for different management strategies of the bridge deck are summarized in Figure 5.11 and demonstrate that a strengthened deck can contribute to a lower LCC under whatever

Table 5.8 *LCC model cost data used for illustrative examples.*

Input parameters	*Details*		*Data*	*Note*
LCC analysis period, T	Designed service life of the structure		100 years	Highway structures are usually designed for 50 to 100 years
Discount rate, r	Convert future costs into present values		4%	Discount Rate. It is recommended to use real discount rates to reflect the true time value of money with no inflation premium. Discount rate at 4% is adopted here for sample illustration only.
Initial design/ construction cost (C_m)*	Carbon steel rebar		HK$ 9,000/ton	Material costs based on bills of quantity of the current on-going highway bridge projects and quotations from major local contractors
	Epoxy-coated rebar		HK$ 12,000/ton	
	Stainless steel rebar		HK$ 35,500/ton	
	Grade 45 concrete		HK$ 1,000/m^3	
	Grade 50 concrete		HK$ 1,450/m^3	
	Waterproof membrane		HK$ 500/m^2	For asphalt overlay
	Protective coating on concrete surface		HK$ 500/m^2	–
	Carbon steel reinforced concrete	Grade 45	HK$ 4,000/m^2	Service life predicted by corrosion models
		Grade 50	HK$ 4,800/m^2	
	Epoxy-coated steel reinforced concrete**	Grade 45	HK$ 9,000/m^2	8 years service life extension to carbon steel reinforced concrete and replacement at end of service life
		Grade 50	HK$ 9,800/m^2	
	Stainless steel reinforced concrete***	Grade 45	HK$15,000/$m^2$	No corrosion-induced damage throughout the service life of the structure
		Grade 50	HK$15,800/$m^2$	

(*Continued*)

Table 5.8 *Continued*

Input parameters	*Details*		*Data*	*Note*
Inspection cost (C_{ins})	General inspection cost ($C_{ins,1}$)		HK$20/m^2	Schedule annually
	Detailed inspection cost ($C_{ins,2}$)		HK$60/m^2	Frequency dependents on service life models. It is assumed that there is no detailed inspection for stainless steel or epoxy-coated reinforced concrete.
Maintenance cost (C_m)	Routine maintenance		HK$40/m^2	Schedule for every 6 months
Replacement and rehabilitation cost (C_{rep})	Do nothing ($C_{rep,1}$)		HK$0/m^2	Do nothing thus, no induced cost
	Minor repair ($C_{rep,2}$)		HK$1,500/m^2	Two-year repair interval frequency
	Major repair ($C_{rep,3}$)	Rebar not corroded	HK$3,000/m^2	Subsequent repair at four-year interval at HK$3,000/m^2
		Rebar corroded	HK$8,000/m^2	
	Rehabilitation ($C_{reps,4}$)	Rebar not corroded	HK$12,000/m^2	Improved durability with approx, 20 years extension of life and subsequent repair at eight-year interval at HK$3,000/m^2
		Rebar corroded	HK$16,000/m^2	
	Replacement ($C_{rep,5}$)		HK$14,000/m^2	Carbon steel RC (maintain same durability)
			HK$16,000/m^2	Epoxy-coated RC (maintain same durability)
Failure cost (C_f)	–		HK$0/m^2	Failure cost is neglected in the model

*Average construction costs per structural unit are estimated based on the rate of rebar and concrete from the bill of quantities of the current ongoing highway bridge projects and quotations from major local contractors as of August of 2008.

** The extension of service life for epoxy-coated steel is based on the assumption that the steel is properly coated and handled throughout the lift of the structure.

*** Stainless steel is not expected to initiate corrosion within 100 years regardless of concrete type.

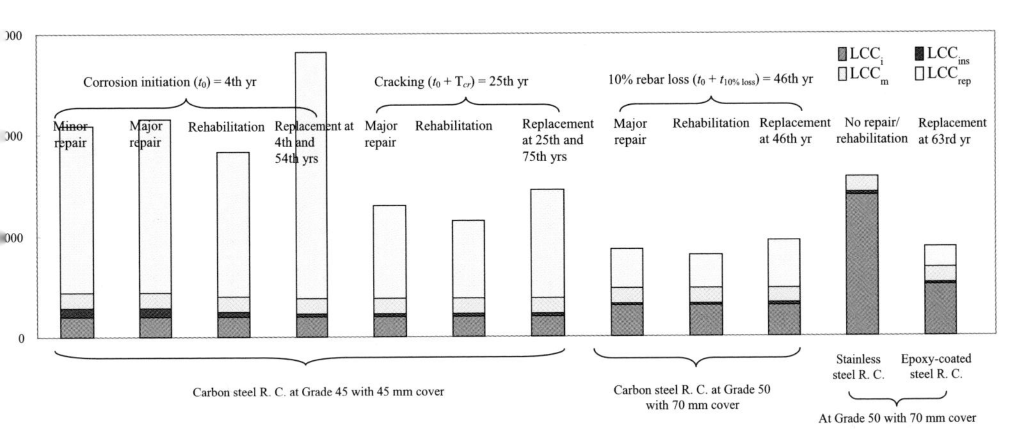

Figure 5.11 *Mean life-cycle cost of different management strategies at bridge deck.*

limit state is adopted (corrosion initiation, severe cracking or 10% rebar section area loss), despite its higher initial design/construction cost (LCC_i). The results also show that the repair & rehabilitation strategy with a service life limit set as the time of corrosion initiation has a higher mean LCC in comparison with the service life limit set at the time of severe cracking.

With reference to Tables 5.5 and 5.6, for the reference case (i.e. Grade 45 with a 45 mm thick cover), as the simulated time to severe cracking ($t_0 + T_{cr}$) is in the 25th year (which is earlier than the time for a 10% rebar section area loss [$t_{10\%\ loss}$]), the proposed dual-management strategy model (Equation 5.40) indicates that option 1, serviceability limit controls the service life.

For the strengthened case (i.e. Grade 50 with a 70 mm thick cover), the simulation results show that a 10% rebar section loss will occur in the 46th year, which is earlier than when severe cracking occurs. With reference to the proposed dual-management strategy model (Equation 5.41) option 2, ultimate limit state controls the service life. It has been found that although its LCC for the initial design/construction is higher, the total mean LCC of the strengthened case is still lower than that of the reference case. It has also been found that the LCC_{rep} of the strengthened case is around 42 to 44% of the LCC_{rep} of the reference case. The comparison results are summarised in Table 5.9.

2. *LCC example 2: Seawater attacks on bridge piers*

The majority of concrete deterioration cases occurring in tropical regions, such as in Hong Kong, is attributed to seawater attack on the splash zone and the tidal zone of bridge piers, particularly for bridges located in coastal areas. The service life of a pier is usually longer than that of a bridge deck as chloride exposure is less severe. Simulation results from figures 5.8 and 5.9 illustrate that whichever service life limit is chosen—corrosion initiation or severe cracking—the splash zone has a shorter service life, so the splash zone always controls the repair & rehabilitation strategy.

The mean LCCs of different management strategies at the bridge pier splash zone are plotted in Figure 5.12. In general, the LCC of repair & rehabilitation work for a pier in the reference case with the service life limit being set at the time of corrosion

Table 5.9 *LCC comparison between base and strengthened cases.*

Location	*Condition*	*Details*	*LCC of R&R works*[a,b]		
			Major repair	*Rehabilitation*	*Replacement*
Deck	Reference case	Grade 45 & 45 mm cover	D_1	D_2	D_3
	Strengthened case	Grade 50 & 70 mm cover	$0.417D_1$	$0.428D_2$	$0.439D_3$
Pier	Reference case	Grade 45 & 45 mm cover	P_1	P_2	P_3
	Strengthened case	Grade 50 & 70 mm cover	$0.360P_1$	$0.373P_2$	$0.342P_3$

[a]D_1, D_2 and D_3 represent different types of LCC_{rep} for the base case of deck.
[b]P_1, P_2 and P_3 represent different types of LCC_{rep} for the base case of pier.

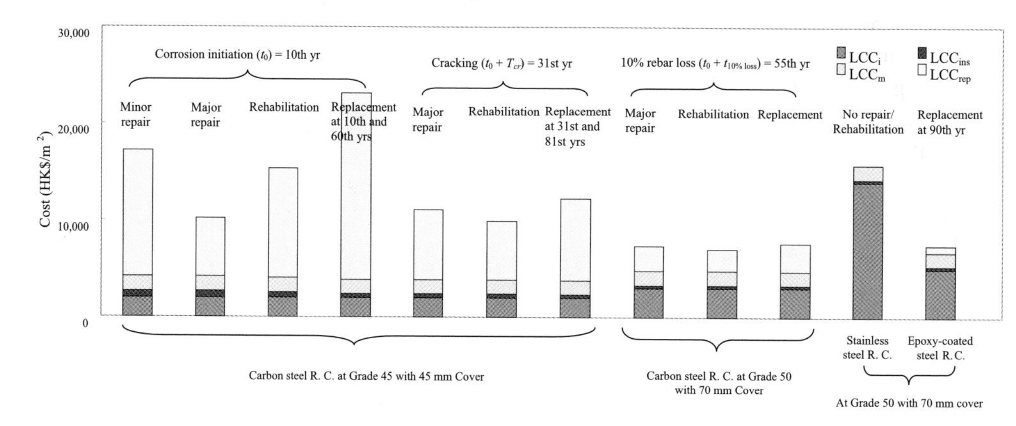

Figure 5.12 *Mean life-cycle cost of different management strategies for a pier.*

initiation (t_0) is higher than the life being set at the time of severe cracking ($t_0 + T_{cr}$), except for a pier under major repair.

As in the deck case, strengthened piers can contribute to a lower LCC, whatever limit state is adopted (corrosion initiation, severe cracking or 10% rebar section area loss), despite its higher LCC_i. The results also show that a repair & rehabilitation strategy with the service life limit set at the time of corrosion initiation has a higher mean LCC in comparison with the service life limit set at the time to severe cracking, except for a pier planned to have major repair work.

Similarly to the deck case, the simulated time for a 10% rebar section area loss for the strengthened case occurs earlier, so the ultimate limit state is the control mechanism. Although its LCC for the initial design/construction is higher, the total mean LCC is still lower than that of the reference case. It has also been found that the LCC_{rep} of the strengthened case is around 34 to 37% of the LCC_{rep} of the reference case. The comparison results are summarised in Table 5.9.

Results and discussions

Recent studies have shown that epoxy-coated steel can give good and long term performance even in severe exposure to chloride conditions and in considering the effects of bond loss when properly coated and handled (Ann and Song, 2007). The service life extension attributable to epoxy-coated rebar in bridge decks was found to be approximately 5 years beyond that of conventional carbon steel bar (Brown *et al.*, 2006). Surely, there is no fixed period of service life extension; in fact, it varies subject to several factors including humidity, temperature, chloride content, workmanship and so on.

Stainless steel reinforcement is employed on a limited basis to address corrosion problems in high-chloride environments. It is not expected to initiate corrosion within 100 years regardless of the concrete type or level of chloride exposure (Williamson *et al.*, 2008). Structures experiencing less severe exposures may benefit from simple improvements in concrete quality and increased cover depth, whereas high risk areas might be more effectively addressed on a case-by-case basis, by using stainless steel reinforcement.

In the previous illustrative LCC analysis examples, it is assumed that epoxy-coated rebar has at least 7 to 8 year service life extension when compared to conventional carbon steel reinforcement with only minimal corrosion induced damaged throughout the life of a bridge (Cui *et al.*, 2007). The purpose is simply to demonstrate a LCC comparison practice for conventional carbon steel reinforcement against other reinforcing materials.

Owing to low initial construction costs, conventional carbon steel is still the most commonly used reinforcing material, rather than epoxy coatings or stainless steel. However, it requires more frequent repair and maintenance throughout its whole life cycle to maintain an acceptable serviceable level. Different types of repair and refurbishment actions and the reaction time are key variables in LCC analysis.

From the examples, it was found that the total LCC varies significantly at different predefined service life limit states. The results demonstrate that it may not always be the case that repair and refurbishment actions at the severe cracking stage produce a lower LCC. Table 5.10 shows that the major repair strategy for the reference case concrete pier with a predefined severe cracking limit results in an even higher LCC.

Table 5.10 *LCC comparison on different serviceability limits in base case.*

Location	*Serviceable limit*	*LCC of R&R works*[a,b]		
		Major	*Rehabilitation*	*Replacement*
Deck	Corrosion initiation	D_1	D_2	D_3
	Severe cracking	$0.54D_1$	$0.53D_2$	$0.44D_3$
Pier	Corrosion initiation	P_1	P_2	P_3
	Severe cracking	$1.19P_1$	$0.53P_2$	$0.44P_3$

[a]D_1, D_2 and D_3 represent LCC_{rep} of different types of R&R works for the base case of deck.
[b]P_1, P_2 and P_3 represent LCC_{rep} of different types of R&R works for the base case of pier.

The results also illustrate that the severe cracking limit state may not always dominate. The strengthened case, as an example, shows that improvement in structure durability could affect the selection of the predefined limit state. Although the strengthened structure has a higher initial design/construction cost, it usually results in a lower LCC.

To conclude, the use of epoxy-coated rebar reinforced concrete bridge deck or pier has the total mean LCC comparable to the strengthened case. The use of a stainless steel reinforcing deck or pier is not justifiable for these two illustrative examples. To select an optimal reinforced concrete management strategy, it is necessary to consider flexibly the selection of the limit state and the different combination of repair and refurbishment actions for particular structural elements under different exposure conditions.

5.5.1.2 LCM Strategy for steel beams of highway bridges

This section presents an integrated LCM strategy considering three structural assessment factors simultaneously: 1) the serviceability limit state of deflection; 2) the ultimate limit state of moment and shear; and 3) the fatigue strength limit state, for steel beams in highway bridges.

A simply supported composite steel beam bridge with rolled-beam stringers is adopted as an example to illustrate the proposed LCM strategy. The bridge has a simple span of 28.0 m with two traffic lanes in the same direction. A 191 mm thick concrete deck slab with a wearing surface of weight 122 kg/m^2 on top is supported by five rolled steel beams of size W36 × 245. Each beam has a welded cover-plate of size 386 × 10 mm thick under a bottom flange. 865 mm high concrete parapets are installed on both sides. The distance from each end of the cover-plate to the adjacent bearing is 2.0 m. The cross section of the bridge is shown in Figure 5.13.

Loading analysis

1. *Dead and superimposed dead*

The dead and superimposed DLs are assumed to be normally distributed with C.O.V. 0.1 (Sommer *et al.*, 1993; Nowak, 1993; and Jiang *et al.*, 2000). The induced mean moments at the mid-span and the end of the cover plate and the induced mean shears at the end supports are calculated and summarized in Table 5.11, in which the maximum bending moment at mid-span is 1,798 kNm, the bending moment at the ends of cover plates is 476.89 kNm and the maximum shear load at the supports is 257 kN for

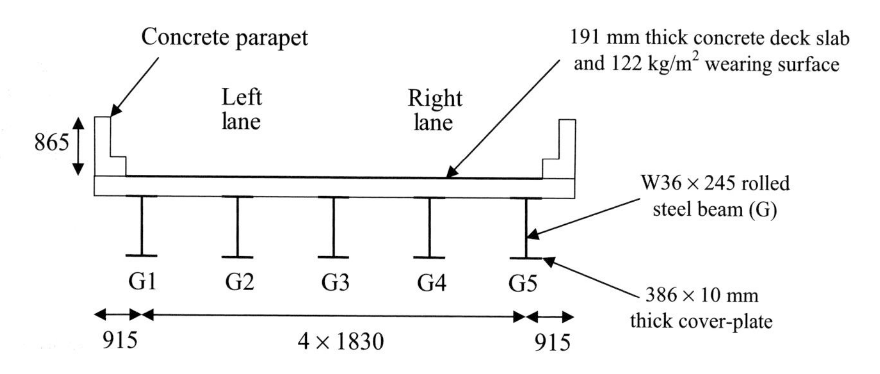

Figure 5.13 *Typical cross section of two-lane rolled steel beam bridge (unit: mm).*

Table 5.11 *Moments and shears due to dead and superimposed dead loads.*

(M_{DL}, M_{SDL}, V_{DL} and V_{SDL})

Beam nos.	*Position*	*Due to dead load*		*Due to superimposed dead*	
		Mean M_{DL} (kNm)	*Mean V_{DL} (kN)*	*Mean M_{SDL} (kNm)*	*Mean V_{SDL} (kN)*
1 and 5	Mid-span	1,185.30	–	612.20	–
	End of cover plate	314.47	–	162.42	–
	Support	–	169.33	–	87.46
2, 3 and 4	Mid-span	1,185.30	–	214.67	–
	End of cover plate	314.47	–	56.95	–
	Support	–	169.33	–	30.67

beams nos. G1 and G5. Whereas, for beam nos. G2, G3 and G4, the maximum bending moment at the mid-span is 1,400kNm, the bending moment at the ends of cover plates is 372 kNm and the maximum shear load at the supports is 200 kN.

2. *Live and impact*

The LL on the bridge is assumed to be due only to traffic. Beam moments and shears summarised in Table 5.12 are calculated based on the survey results of Nowak (1993), with literature references in which the average daily truck traffic (ADTT) is 5,000, with 66% of the cases having a truck in the left lane, 33% of the cases having a truck in the right lane and 1% of the cases having trucks in both lanes (Sommer *et al.*, 1993; Jiang *et al.*, 2000; Cheung and Li, 2003). It is also assumed that the LL from the vehicles is kept unchanged throughout the bridge's service life. To account for the dynamic effects of a vehicle riding over the bridge, an impact factor is adopted as a multiplier. The impact fraction of the LL according to the AASHTO Specification (2004) is 0.231. The moment and the shear due to impact will then be added to the LL schedule.

Service life prediction

1. *Serviceability limit state: deflection*

Corrosion reduces the effective cross-section area and occurs mainly at the web and the top of the bottom flange of steel girders. Reduction of section area leads to weakening of the flexural strength and the shear strength, and increases the deflection of the bridge deck under service loadings. According to AASHTO specifications, the deflection limit due to the LL plus impact is span/800 = 35 mm for typical highway bridges. In order to investigate the long term effectiveness of different preventive measures for steel girders, three scenarios, a) carbon steel without protective paint, b) carbon steel with protective paint and c) weathering steel, are compared. The steel bridge is modelled against corrosion in marine environments based on the statistical parameters listed in Table 5.13.

Table 5.12 *Moments and shears due to live load (M_{LL} and V_{LL}).*

Beam no.	*Position*	*Left lane*			*Right lane*			*Both lanes*		
		Mean M_{LL} (kNm)	*Mean V_{LL} (kN)*	*C.O.V.*[(1)]	*Mean M_{LL} (kNm)*	*Mean V_{LL} (kN)*	*C.O.V.*[(1)]	*Mean M_{LL} (kNm)*	*Mean V_{LL} (kN)*	*C.O.V.*[(1)]
1	Mid-span	2,333.46	–	0.24	567.56	–	0.33	2,720.61	–	0.23
1	End of cover plate	815.24	–		198.29	–		950.50	–	
1	Support	–	333.35		–	81.08		–	388.66	
2	Mid-span	2,390.08	–	0.18	998.17	–	0.23	3,247.35	–	0.18
2	End of cover plate	835.02	–		348.73	–		1,134.52	–	
2	Support	–	341.44		–	142.6		–	463.91	
3	Mid-span	1,786.97	–	0.19	1,735.61	–	0.19	3,356.65	–	0.19
3	End of cover plate	624.31	–		606.37	–		1,172.71	–	
3	Support	–	255.28		–	247.94		–	479.52	
4	Mid-span	1,107.47	–	0.23	2,413.79	–	0.18	3,356.65	–	0.19
4	End of cover plate	386.92	–		843.30	–		1,172.71	–	
4	Support	–	158.21		–	344.84		–	479.52	
5	Mid-span	654.47	–	0.27	2,440.12	–	0.22	2,920.77	–	0.21
5	End of cover plate	228.65	–		852.50	–		1,020.43	–	
5	Support	–	93.50		–	348.59		–	417.25	

[a]C.O.V.: Coefficient of variation.

Table 5.13 *Statistical parameters of random variables for the steel beam bridge example.*

Parameters	*Materials*	*Variables*	*Mean (μ)*	*C.O.V.*[(1)] *(σ/μ)*	*Distribution*	*Sources*
Corrosion deterioration rate	Carbon steel	A	70.6 μm	0.66	LN[(2)]	Albrecht and Naeemi, 1984; Kayser, 1988
		B	0.789	0.49		
		ρ_{AB}	–0.31	–	–	
	Weathering steel	A	40.2 μm	0.22	LN	
		B	0.557	0.10		
		ρ_{AB}	–0.45	–	–	
Corrosion initiation	–	T_{CI}	15 yr	0.30	LN	Lee *et al.*, 2006
Repaint duration	–	T_{REP}	20 yr	0.25		
Initial crack dimension	Structural steel	a_0	0.762 mm	0.5	LN	Cheung and Li, 2003
Crack growth constant		C	1.26×10^{-13}	0.63		
Crack growth exponent		m	3	–	Constant	Fisher, 1984
Critical crack dimension		a_c	0.0254 m			
Fracture toughness		K_{IC}	43.97 MPa $m^{0.5}$	0.19	TN[(3)]	Jiang *et al.*, 2000
Compressive strength	Concrete	f_c	21 MPa[(4)]	0.19	LN	Sommer *et al.*, 1993
Yield stress	Steel	F_y	248 MPa	0.10		
Modulus of elasticity	Concrete	E_c	30,000 MPa	0.20	LN	Cheung and Li, 2001
	Steel	E_s	210,000 MPa	0.06		
Deck slab thickness	Concrete	t_c	191 mm	0.20	N[(4)]	Cheung and Li, 2003

[a]C.O.V. : Coefficient of variation.
[b]LN: Lognormal distribution.
[c]TN: Truncated normal distribution.
[d]N: Normal distribution.

The nominal maximum deflections under the three different scenarios are calculated according to equations 5.13 to 5.16 and summarized in Table 5.14. The deflection limit against time over the life span is plotted in Figure 5.14. The results show that the deflection for girders without any protective system is the highest, while the deflection for girders made of weathering steel is the lowest. According to the proposed management strategy, throughout the 75-year service life, none of steel beam types falls below the fair condition state (i.e. 'F' – state) in the deflection condition state set (i.e. Equation 5.47). In other words, no repair and rehabilitation work for deflection is expected throughout the lifespan.

Table 5.14 *Deflection of steel girder bridge with cover plate.*

Years	*Nominal corrosion penetration (mm)*			*Nominal maximum deflection (mm)*		
	Carbon steel without coat	*Carbon steel with coat*	*Weathering steel*	*Carbon steel without coat*	*Carbon steel with coat*	*Weathering steel*
0	0.000	0.000	0.000	29.332	29.332	29.332
10	0.434	0.000	0.145	29.554	29.332	29.405
20	0.750	0.251	0.213	29.719	29.460	29.440
30	1.033	0.251	0.267	29.870	29.460	29.468
40	1.297	0.503	0.314	30.012	29.589	29.492
50	1.546	0.503	0.355	30.149	29.589	29.513
60	1.786	0.754	0.393	30.282	29.721	29.533
70	2.016	0.754	0.428	30.413	29.721	29.551
75	2.129	0.754	0.462	30.478	29.721	29.560

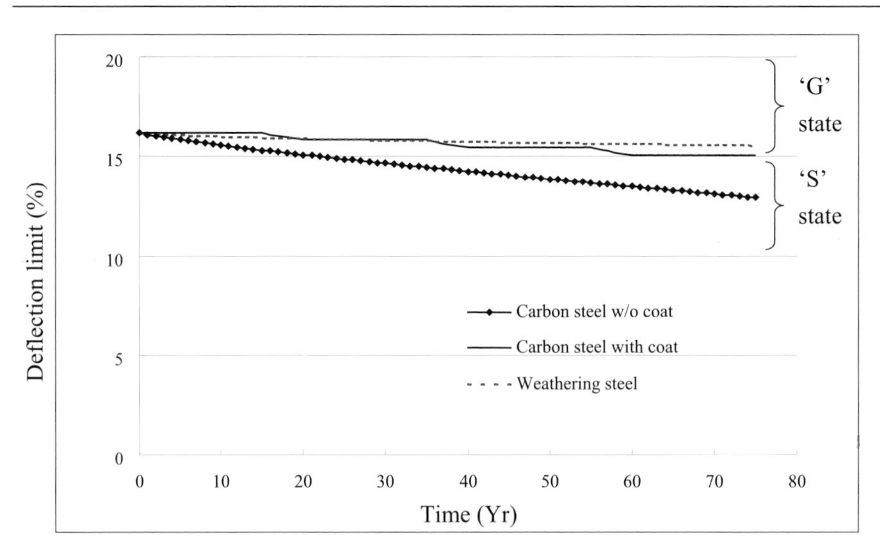

Figure 5.14 *Deflection limit over lifespan.*

2. *Ultimate limits: moment and shear*

It is assumed that R&R actions take place only when the condition reaches the 'P' state range in accordance with equations 5.48 and 5.49. Simulated times to reach the 'P' state in terms of moment and shear limits at the 80 percentile value are plotted in figures 5.15 and 5.16, respectively. Concerning the moment condition, the result shows that beam no. G1 will reach the moment limit in the 26th year and the other beams will remain in the 'F' state and 'S' state throughout their lifespan. As to the shear condition, results demonstrate that all beams will remain in the 'G' state, which means shear stress has insignificant impact on the beams.

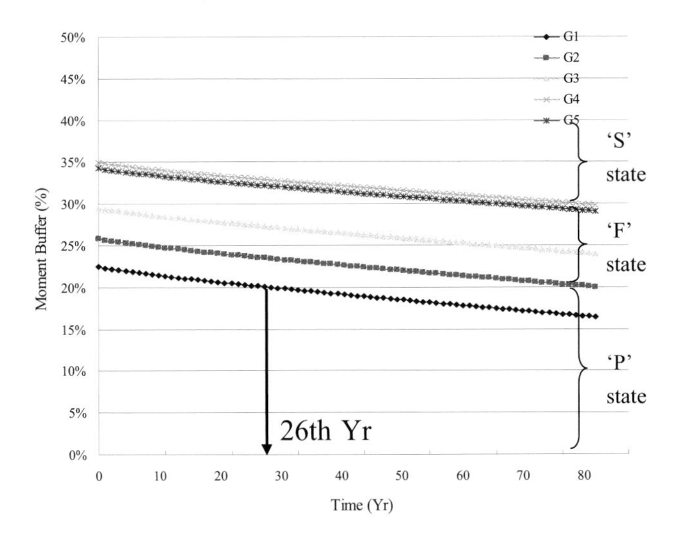

Figure 5.15 *Moment limit over lifespan of carbon steel beams.*

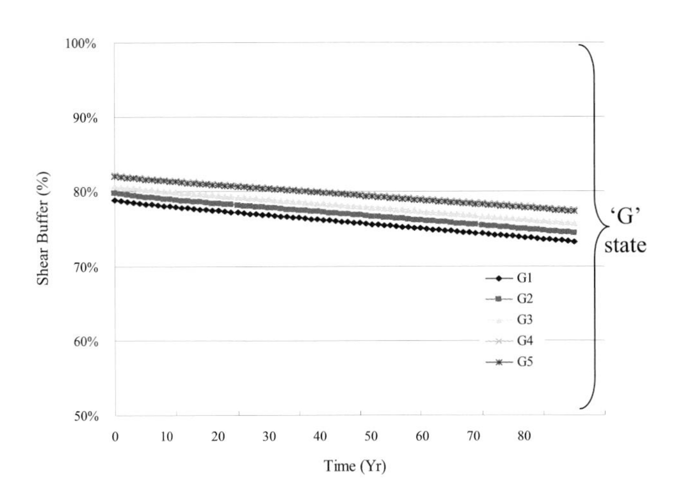

Figure 5.16 *Shear limit over lifespan of carbon steel girders.*

LIVERPOOL JOHN MOORES UNIVERSITY
LEARNING SERVICES

3. *Fatigue damage limit state*

The maximum LL model afore-developed is insufficient to determine the effective range of the stress intensity factors. For fatigue analysis, the loading effects are modelled by the fatigue truck provided in the ASSTHTO specification, with gross weight 240 kN. Each truck passage is assumed to cause only one stress cycle. Under repeated tensile stresses, fatigue cracks may form at the end weld of the cover plate and penetrate into the bottom flange. Under a given bending moment in a beam, the critical fatigue stress is located in the bottom flange at the end of cover plate such that it should be calculated based on the cross section of the beam without the cover plate. The service life of each beam is simulated based on the statistical parameters in Table 5.13, and the simulated time for each beam to reach its 'P' state is summarized in Table 5.15. The results show that beam no. G2 is the most sensitive for fatigue damage and will reach the fatigue limit in the 43rd year.

Results and discussions

Clearly, not only moment and shear capacities are affected by corrosion, but also bearing capacity. However, in common practice, if a steel bridge does not receive proper maintenance and painting, it will probably be constructed with bearing stiffeners to increase the bearing capacity.

In order to reduce the complexity in showing the implementation of the proposed integrated LCM strategy, a technical analysis of bearing behaviour is not included in the study. A steel beam bridge is provided as an example to demonstrate the use of the proposed strategy with corrosion deterioration and fatigue damage models being incorporated.

There are two important messages incurred from the example: 1) Fatigue damage may not always dominate. Conventional LCM is usually in accordance with fixed predefined limit states, either the serviceability limit state, the ultimate limit state or the fatigue limit state. However, it may not always be valid that one limit state dominates over another in all situations.

The results demonstrate that 1) the earliest simulated replacement or rehabilitation time for beams under bending stress is in the 26th year while the earliest simulated repair time for beams under fatigue damage is in the 43rd year; and 2) the service lives of each beam are not constant. The conventional management approach treats all beams as equal. However, some beams may deteriorate faster than expected. In the case of fatigue damage, the longest life is for beam no. G5 while the shortest is beam no. G2. Concerning the bending failure, the most critical is beam no. G1, while the least critical are nos. G4 and G5.

Table 5.15 *Time (year) to 'P' state of fatigue limits.*

	Beam no.				
	G1	*G2*	*G3*	*G4*	*G5*
Time	48	43	47	50	62

These service life prediction models integrating management strategy can provide a better picture for the stakeholders to identify any sensitive steel beams under corrosion deterioration or fatigue damage. Other factors affecting the condition of steel beams are neglected in the study. However, steel component damage is definitely not exclusive to these two types.

Moreover, proposed acceptable limits can be further tied into the reliability index if sufficient statistical information of the parameters in the deterioration models is available. The rehabilitation time can be determined by the predefined reliability index value. Even though the reliability assessment has not been incorporated in the study, the proposed LCM concept is not affected.

Furthermore, owing to the lack of sufficient cost data, the study only addresses the LCM strategy and the methodology of service life prediction models being integrated into the LCC model. If sufficient financial data is available, it could provide further decision making information for selecting the most appropriate management strategy by LCC comparison.

Therefore, further research approaches on the proposed LCM strategy for steel beam bridges can be conducted as shown in Figure 5.7, to test and verify its practicability. The proposed verification approach is categorized into 7 stages from the design stage to the analysis stage. However, some key data should be obtained in the investigation stage prior to the LCC analysis process.

- *Cost data*

Collection of possible future cost data is necessary to provide further decision making information for selecting the most appropriate management strategy. The cost of different types of future work should comprise inspection, operation, maintenance, management, repair, rehabilitation, replacement, demolition and failure.

- *Historical data*

Collection of historical repair or rehabilitation data can be used to estimate the probability of any occurrence in case the future action time could not be predicted by appropriate deterioration models.

- *Deterioration mechanism*

Steel component damage is definitely not exclusive to corrosion deterioration and fatigue damage. In the study, other factors affecting the condition of the steel beams are neglected. In fact, the bearing capacity of the steel beams and deterioration of other bridge components such as bridge bearings, expansion joints, profile barriers and so on could be incorporated into the LCM strategy. The proposed LCM strategy can be further enlarged.

5.5.2 Application in public-private partnerships procurement approach

Why has there been such a worldwide growth in interest in public-private partnerships (PPPs) over the last decade? One of the main drivers for growth is that PPPs could avoid limitations on public-sector budgets. However, the detailed debate on the merits of PPPs is highly complex.

The idea of PPPs, as now commonly known, came about in the United Kingdom (UK) in the 1990s, when many government projects were running late and over budget, and the UK Government was keen to reduce public sector borrowing. PPPs is an attractive proposition because it not only allows a government to pass the financial burden onto the private sector, but it also promises to deliver better projects because the way a PPP project is structured encourages the concessionaire to think about long-term issues such as operation and maintenance costs (Kwan, 2005).

PPPs are based on a partnership approach, where the responsibility for the delivery of services is shared between the public and the private sector, both of which bring their complementary skills to the enterprise. In other words, the public sector will engage private sector providers to develop facilities and/or deliver services that private sector can provide more effectively and efficiently (Ho, 2005).

According to the Efficiency Unit of the Government of the HKSAR (2003), PPPs are 'arrangements where the public and private sectors both bring their complementary skills to a project, with varying levels of involvement and responsibility, for the purpose of providing public services or projects'.

In fact, a PPP is a medium to long-term relationship between the public and private sectors including the voluntary and community sectors, involving the sharing of risks and rewards and the utilization of multisectoral skills, expertise and finance to deliver desired policy outcomes or projects that are in the public interest.

5.5.2.1 Integration of life-cycle management approach

LCM is perhaps the most important element of the value for money case for PPPs. What is value for money? It is the combination of risk transfer, whole LCC and service provided, as a basis for deciding what offers the best value to the public sector entity. Because the same investors will usually be responsible for both the construction of the facilities and for its operation and service delivery, they are incentivized to design it to produce the best LCC. For example, private-sector investors may be prepared to spend more on the initial capital cost if this will result in a greater saving in maintenance costs over the life of the PPP contract, whereas a typical public-sector procurement approach is to go for the lowest initial capital cost (Yescombe, 2007).

Furthermore, the significance of the risk-transfer argument should not be neglected. A PPP transfers the maintenance-cost risk—probably the most difficult to predict—to the private sector. Having capital at risk ensures that the investor in and lender to the project company cannot easily walk away from this risk.

It can also be said that the long term contractual nature of PPPs forces the public-sector to make provision for maintenance (through the service fees), without regard to short-term budget constraints which might otherwise encourage the omission of routine maintenance, and at the same time incentivizes the private sector to carry out the maintenance if service fees are not paid (or deductions made) when maintenance standards are not met. A PPP contract thus should ensure facilities such as highway structures, are maintained to predetermined standards, throughout the life. However, a public authority could enter into a long-term contract covering design construction and maintenance producing the same result.

5.5.2.2 Public-private partnerships in Hong Kong

In Hong Kong, PPP was first mentioned several years ago, when the government accumulated a budget deficit following the Asian financial crisis and had to explore ways to cut expenditure and still deliver much-needed infrastructure. Since then, several projects have been proposed for consideration, sparking much debate as to whether PPP is the appropriate model for infrastructure delivery in Hong Kong (Kwan, 2005).

PPPs cover a spectrum of contractual approaches between the government and the private sector, involving complex projects, large sums of money and long-term arrangements. In the Hong Kong context, the BOT approach is used in the provision of major road tunnels, and the DBO (design-build-operate) approach is used in the introduction of sophisticated solid waste management facilities. Recent major PPP projects include the Disney theme park, the AsiaWorld-Expo located at the Hong Kong International Airport, the Tung Chung Cable Car connecting the railway network to the bronze statue of Buddha, to name but a few. Other possible major PPP projects coming on stream include the reprovisioning and operation of the Shatin Water Treatment Work and the redevelopment of the Central Police Station.

While many associate PPP with cost saving, the Hong Kong government does not implement PPPs primarily as a way to balance the books, as PPPs should not be used as a quick-fix to budgetary and staffing constraints. The government considers it worth infusing the entrepreneurial spirit of the private sector into public service, thereby enhancing productivity and flexibility, as well as capitalising on the skills and experience, that are not available in-house. The management skills and financial acumen of the business community would hopefully provide the government with greater value for money.

With respect to the operation stages of highway infrastructure, the conventional mode of highway maintenance usually follows a deterministic approach that is a highly labour intensive for site inspection, estimating, checking, measuring and so forth. Whereas, the newly introduced PPP procurement approach helps to alleviate the government's financial burden and transfers the project risks properly to the private sector entities. With the integration of the LCC model, the financing strategy and the project period can be better developed and estimated.

5.5.2.3 Merits of public-private partnerships

PPP can take different forms to meet the specific needs of individual cases. The following is a summary of the potential benefits of the implementation of PPP including some Hong Kong cases that have been quoted for elaboration.

1. Allowing government's strategic development

   The PPP approach is appealing to government because it can then concentrate on its core competencies by minimizing its capital costs under a limited budget and consistent with its management philosophy—maintaining a small government and lean civil service. Among PPP models, the BOT/DBOM (design, build, operate and maintain) models combining the design and construction responsibilities of design-build procurements with operation and maintenance are most often used in Hong Kong.

   In fact, this project delivery approach is practised by several governments around the world. These models transfer design, construction and operation of a single facility or group of assets to a private sector entity. The government can take advantage of operational efficiencies regularly associated with private sector participation and the enhanced unity of responsibilities for delivering services

2. Enhancing social development and business opportunities
   The PPP approach alleviates the burden, both financial and managerial, and creates economic growth, employment and investment opportunities. The private sector entities could be brought into the provision of public services such that the social and public services could be developed sustainably without inducing a great burden on the government. PPP projects focusing on the leisure and cultural facilities such as the Ice Sports Centre in Area 45, Tseung Kwan O, the Leisure and Cultural Centre in Kwun Tong and the controversial project—West Kowloon Cultural District—are examples which demonstrate that the involvement of private sector entities could be of potential benefit to society. Another PPP project, Cyberport, is an example for the government to generate new sources of revenue.
3. Encouraging technology advancement
   Assets or facilities operation and maintenance responsibilities in PPP arrangements are usually the responsibility of the private sector entities. The performance standards and specifications for the operational and maintenance assets and facilities are stated in the contract documents. To maximize profit, one of the major approaches is better financial control of the LCC. This initiative invokes the private sector skills and experiences in accessing new technology and innovation with a view to minimizing the operating and maintenance costs. As a result, the private sector entities would rather invest and introduce new or improved technology for better delivery of public services.
4. Better risks allocation
   One of the major initiatives in implementing PPP can result in less risk of cost overruns and project delays. In a longer contract time frame, the resulting uncertainties, variables, unknowns, variety of organisations and people involved induce large risks. The private sector will have a stronger incentive and presumably a greater capacity to manage these risks efficiently. The risks are high, but so are the potential rewards. Risk mitigation measures can be formulated and incorporated in decision-making strategies and procedures at all stages of the PPP project, no matter whether evaluating the feasibility of PPP, selecting the precise form of PPP, selecting the concessionaire/franchisee or formulating the design and construction checking procedures.
5. Enhancing incorporation of life-cycle management
   Projects under PPP arrangements have a much greater emphasis on whole life cycle performance in the determination of capital expenditure. The whole life cycle performance of a project is dependent upon the whole life costs, life cycle assessment, quality and public acceptance. From the government point of view, cutting spending or delivering infrastructure without additional government expenditure or spreading government's capital investment over the life of a project are the major concerns. From the private sector entity point of view, minimizing the maintenance cost is one of the ways of profit making. Indeed,

in the design stage, a private sector entity would undertake LCC analysis of the facilities or assets such that a cost model could be developed to suit the performance standards specified in the PPP contract documents. In addition, PPP could also provide better exploitation of government assets and intellectual property.

6. Enhancing sustainability
   There has been a tendency in the past for the contractor to finish the project in the cheapest and fastest way, an approach that often compromises quality and leaves the owners with high recurrent maintenance and operational costs. The PPP approach motivates the private partner to design and construct facilities to a high standard, and to focus more on the operational efficiency, aiming for an overall reduction in the LCC, especially related to energy consumption. Taking building assets as examples, minimizing the operating cost is taken care of by attention to building services, as well as architectural design. Careful zoning contributes to low electricity consumption by giving the operator the flexibility to operate building services equipment as needed.
7. Quality improvement
   The effectiveness and improvement of assets or facilities increase with private sector involvement. The private sector has acquired a good deal of financial and technical experience in many areas, and is also interested in upgrading the service level, which satisfies the public interest. The government and controlling bodies may benchmark the performance and quality of services provided by the private sector. This type of benchmark can be used to measure the quality of services provided by government agencies. Some forms of private involvement can be used as mechanisms for structural reform, as in the case of economic facilities management contracts. In more enlightened cases, there is a real recognition that the private sector possesses management expertise and innovative ideas that will help operate the facilities or services better than the public sector.

5.5.2.4 Forms of public-private partnerships

Quite a number of development or service delivery models come under the PPP umbrella. With reference to the information shown on the websites of the U.S. Government Accounting Office and the Canadian Council for public private partnerships (PPP) can be established in the following form and sequence with regard to their degree of private sector involvement, namely:

1. Design and build (DB)
   The private sector designs and builds infrastructure to meet public sector performance specifications, often for a fixed price. As a result the risk of cost overruns is transferred to the private sector. One important thing is that the public sector still owns the assets and has the responsibility for the operation and maintenance.
2. Maintenance, operation and management services (MOM)
   A private operator, under contract, operates, maintains and manages a publicly-owned asset, facility or system providing a service for a specified

term. Ownership of the asset remains with the public entity. In general, the longer the contract term, the greater the opportunity for increased private investment since more time is available for the private operator to earn profits. Through competitive bidding on the contract, the public can benefit from a reduction in delivery costs and improved service quality.

3. Operation licence
   A private operator receives a licence or rights to operate a public service, usually for a specified term. This is often used in IT projects rather than infrastructure construction projects.
4. Finance only
   A private sector entity, usually a financial services company, funds a project directly or uses various mechanisms such as a long-term lease or bond issue.
5. Design, build and maintenance (DBM)
   Similar to a DB contract except the maintenance of a facility becomes the responsibility of the private sector entity.
6. Design, build and operate (DBO)
   A single contract is awarded for the design, construction and operation. DBO allows continuity of the private sector involvement. It facilitates private sector entity financing in the form of user fees.
7. Design, build, finance and operate (DBFO)
   DBFO concerns the private sector itself handling the design, build and financing of the project. DBFO is also known as private finance initiative (PFI) in the UK. The public sector has to pay an annual fee to the private sector entity under a long-term operating contract for the service.
8. Build, operate and transfer (BOT)/Build, transfer and operate (BTO)
   A private sector entity receives a contract or franchise agreement to finance, design, build and operate a facility for a specified time period, after which ownership is transferred back to the public sector. The private sector entity usually provides all or part of the financing, and the contract is structured to be of sufficient length to enable the private sector entity to realize a reasonable profit.

   The BTO model is similar to the BOT model except that the transfer of the facility to the public agency takes place at the time that construction is completed, instead of at the end of the franchise period. In Canada, BOT is usually called Build, own, operate and transfer (BOOT).
9. Build, own and operate (BOO)
   BOO is used to describe public facilities or services which are funded, built, owned and run by the private sector entity for a fixed number of years. Under this type of model, the private sector entities do not need to transfer the ownership of the facility to the public sector such that they can maximize their return on the investment. The public constraints are stated in the original agreement and through the on-going regulatory authority. While this approach is more common in the power and telecommunications sectors, it has also been used to develop transportation infrastructure.

10. Build, develop and operate (BDO)
    A private sector entity leases or buys an existing facility from a public agency, invests its own capital to renovate, modernize or expand the facility and then operates it under a contract with the public agency
11. Buy, build and operate (BBO)
    This arrangement is a form of public assets and facilities transfer to a private or quasipublic entity. Usually, under this contract, the assets and facilities are to be rehabilitated, upgraded, expanded and operated for a specified period of time.

5.6 Conclusion

Highway bridges should be maintained to remain safe and functional during their service lives to enable personal mobility and transport of goods to support the economy and ensure high quality of life. Stakeholders including maintenance authorities and pubic users usually expect their bridges to have a service life of 50 to 100 years, with only routine maintenance. In fact, demands on most bridges have been increasing annually because of growing traffic volumes, higher loads and aggressive environments. These conditions, coupled with the inadequate funding allocated for maintenance, have led to the accelerated aging and extensive deterioration of these critical structures. The consequences of a bridge failure can vary from a minor disruption of traffic to catastrophic collapse with injuries and loss of life. Therefore, it is imperative that rigorous and reliable models and techniques are used to predict the performance and assess the corresponding risk of failure throughout the service life of bridges (Lounis and Daigle, 2007).

The growing concerns with climate change and the demands for the reduction of greenhouse gas emissions, energy efficiency and conservation of raw materials present considerable challenges to owners, engineers and all construction industries. To address these challenges, there is a need to develop effective approaches for LCM of highway bridges that will ensure their sustainability over a long planning horizon, in terms of improved physical performance, cost-effectiveness and environmental compatibility (*ibid.*).

These optimized designs and management systems should provide the owners with the solutions that achieve an optimal balance among three relevant and competing criteria, namely: (1) engineering performance (e.g. safety, serviceability and durability); (2) economic performance (e.g. minimum LCC and minimum user costs); and (3) environmental performance (e.g. minimum greenhouse gas emissions, reduced materials consumption, energy efficiency and so forth).

Owing to their scopes, the proposed integrated LCM approach for highway bridges only takes into account their first two performance criteria, engineering and economic, over their life cycle, with emphasis on the service life prediction and reduction of LCC. The implementation of this approach requires prior knowledge of the service life of highway bridges for different design and maintenance alternatives in order to select an effective design and associated maintenance strategy that will minimize the LCC of the bridge system. Therefore, a reliable prediction of the service life of a highway bridge system built in a given environment and subjected to different environmental and mechanical loads is of utmost importance in order to obtain reliable estimates of its LCC.

The increasing numbers of highway structures and the deterioration of infrastructure systems present great financial, safety, technical and operational challenges to governmental organisations in charge of public infrastructure, development and management (So *et al.*, 2009). In order to facilitate efficient and effective infrastructure facilities management, this research study describes the development of integrated performance-based LCM strategies for highway bridges.

Intensive literature reviews have been conducted to understand better the recent development in LCM research. The concepts of LCM adopted in this chapter of different management considerations at various project stages are summarized. Fundamentals of LCC and the establishment of LCC models for highway bridges are addressed. The methodology on how service life prediction models are being integrated into the LCM strategies are developed

Some previous works in deterioration and structural damage models that affect the ultimate strength of structures have also been assessed. Service life prediction models for concrete relative to corrosion deterioration and structural steel elements relative to fatigue damage are also covered. The study also proposes practical management strategies and develops performance-based LCC management model for:

1. Reinforced concrete structures relative to corrosion deterioration
   The corrosion mechanism of reinforced concrete structures, and corrosion models, have been adopted to predict the predefined limit states through a probabilistic method, namely MCS. Continuous and systematic maintenance of highway structures can extend the service life and reduce operating expenses. A comprehensive LCM framework is proposed covering inspection, maintenance and repair strategies for maintaining and continuing adequate serviceable levels of safety and convenience to road users. Possible inspection, maintenance and R&R action strategies and their cost data are also established.
2. Steel bridge beams relative to fatigue damages
   This chapter also presents a proposed steel beam bridge management framework assisting stakeholders to appropriately and reasonably prioritize their future maintenance related work for their bridge stock such that stakeholders can allocate better their limited resources. An illustrative example on the performance-based LCM strategy on steel bridge beams is provided to demonstrate its applicability for highway bridges.

Finally, an overview on the applications of LCM particularly in the contracting and project financing stages is conducted and the benefits of an integrated LCM strategy into public-private partnership procurement approach are elaborated.

To conclude, this chapter presents a practical concept for assisting stakeholders to choose appropriately the best preventive measures for long-term maintenance or treatment for bridge rehabilitation affected by chloride-induced corrosion deterioration and fatigue damage. The decision analysis is referred to the whole LCC approach by considering appropriate elements of bridge rehabilitation costs. This practical framework for the LCM strategy is presented to understand better and apply the methodology for life-cycle design and provide a guide for the design of new bridges and the assessment and management of existing highway bridges.

Acknowledgment

The book chapter preparation works was financially supported by the Seed Grant of the Technological and Higher Education Institute of Hong Kong.

Bibliography

AASHTO (2004) *AASHTO LRFD bridge design specifications*. American Association of State Highway and Transportation Officials, Washington, D.C.

AASHTO (2008) *Bridging the gap: restoring and rebuilding the nation's bridges*, American Association of State Highway and Transportation Officials, Washington, D.C.

Albrecht, P. and A.H. Naeemi (1984) Performance of weathering steel in bridges, *National Cooperative Highway Research Program Report 272*. Transportation Research Board, National Research Council, Washington, D.C.

Ang, A. and W. Tang (1975) *Probability concepts in engineering planning and design: Volume I—Basic principles*. Wiley, New York, NY.

Ann, K.Y. and H.W. Song (2007) Chloride threshold level for corrosion of steel in concrete. *Corrosion Science* **49**, 4113–4133.

Ashworth, A. and K. Hogg (2000) *Added value in design and construction*. Longman Publishing, London, UK.

Bazant, Z.P. (1979) Physical model for steel corrosion in concrete sea structures—applications. *Journal of Structural Division* **105**(6), 1155–1165.

Bentur, A., S. Diamond and S. Berken (1997) *Steel corrosion in concrete: fundamentals and civil engineering practice*, 1st ed. E & FN Spon, London, UK.

Bhaskaran, B., N. Palaniswamy and N.S. Rengaswamy (2006) Life-cycle cost analysis of a concrete road bridge across open sea. *Materials Performance* **45**(10), October, 51–55.

Biondini, F., F. Bontempi, D.M. Frangopol and P.G. Malerba (2006) Probabilistic service life assessment and maintenance planning of concrete structures. *Journal of Structural Engineering* **132**(5), May 1, 810–825.

Boussabaine, A. and R. Kirkham (2004) *Whole life-cycle costing: risk and risk responses*. Blackwell Publishing, Oxford, UK.

Branco, F.A. and J. De Brito (2004) *Handbook of concrete bridge management*. American Society of Civil Engineers, Reston, VA.

British Standards Institution (1985) BS 8110. Structural use of concrete—part 1: Code of practice for design and construction. BSI, London.

British Standards Institution (2000) BS EN 206-1. Concrete: Specification, performance, production and conformity. BSI, London.

Broomfield, J.P. (1997) *Corrosion of steel in concrete: understanding, investigation and repair*, 1st Edition. E & FN Spon, London, UK.

Brown, M.C., R.E. Weyers and M.M. Sprinkel (2006) Service life extension of Virginia bridge decks afforded by epoxy-coated reinforcement, *Journal of ASTM International* **3**(2), February, 1–13.

Bull J. (1993) *Life cycle costing for construction*. Chapman & Hall, London, UK.

Cady, P.D. and R.E. Weyers (1983) Deterioration rates of concrete bridge deck. *Journal of Transportation Engineering*, ASCE **110**(1), 34–44.

Caner, A., A.M. Yanmaz, A. Yakut, O. Avsar and T. Yilmaz (2008) Service life assessment of existing highway bridges with no planned regular inspection. *Journal of Performance of Constructed Facilities* **22**(2), 108–114.

Chen, D. and S. Mahadevan (2008) Chloride-induced reinforcement corrosion and concrete cracking simulation. *Cement & Concrete Composite* **30**(3), 227–238.

Cheung, M.M.S., X.Q. Zhang and K.K.L. So (2006) Life cycle cost analysis and management of reinforced/prestressed concrete structures. *Proceedings of the 5th International*

Workshop on Life-Cycle Cost Analysis and Design of Civil Infrastructure Systems, Seoul, South Korea, 11–14 October.
Cheung, M.M.S., J. Zhao and Y.B. Chan (2009) Service life prediction of RC bridge structures exposed to chloride environments. *Journal of Bridge Engineering* **14**(3), May 1, 164 –178.
Cheung, M.S. and B.R. Kyle (1996) Service life prediction of concrete structures by reliability analysis. *Construction and Building Materials* **10**(1), 45–55.
Cheung, M.S. and W.C. Li (2001) Serviceability reliability of corroded steel bridges. *Canadian Journal of Civil Engineering* **28**, 419–424.
Cheung, M.S. and W.C. Li (2003) Probabilistic fatigue and fracture analyses of steel bridges. *Structural Safety* **23**, 245–262.
County Surveyors' Society (2004) *Framework for highway asset management*. County Surveyors' Society, London, UK.
Crank, J. (1975) *The mathematics of diffusion*, 2nd ed. Oxford University Press, Oxford, UK.
Cui, F., J. Lawler, P. Krauss and Wiss, Janney, Elstner Associates, Inc (2007) Corrosion performance of epoxy-coated reinforcing bars in a bridge substructure in marine environment. *Corrosion 2007*, March 11–15, Nashville, Tennessee.
Czarnecki, A.A. and A.S. Nowak (2008) Time-variant reliability profiles for steel girder bridges. *Structural Safety* **30**, 49–64.
Dale, S.J. (1993) Introduction to life cycle costing. In: *Life Cycle Costing for Construction*, J.W. Bull, ed. 1st ed. Chapman & Hall, London, UK, 1–22.
Darmawan, M.S. (2010) Pitting corrosion model for reinforced concrete structures in a chloride environment. *Magazine of Concrete Research* **62**(2), February, 91–101.
DuraCrete (2000) Project No. 95 – BE1347: *Probabilistic performance based durability design of concrete structures*. The European Union-Brite EuRam III.
Efficiency Unit (2008) *Serving the community by using the private sector: an introductory guide to public private partnerships (PPPs)*, 2nd Edition. Efficiency Unit, HKSAR Government, March.
Eibl, J. (ed.) (2004) *Concrete structures Euro-design handbook: 1994/96*. Ernst & Sohn, Berlin, Germany.
El Maaddawy, T. and K. Soudki (2007) A model for prediction of time from corrosion initiation to corrosion cracking. *Cement & Concrete Composites* **29**(3), 168–175.
Enright, M.P. and D.M. Frangopol (1998) Probabilistic analysis of resistance degradation of reinforced concrete bridge beams under corrosion. *Engineering Structures* **20**(11), 90–97.
Fisher, J.W. (1984) *Fatigue and fracture in steel bridges: case studies*. John Wiley & Sons, New York, NY.
Frangopol, D.M., K-Y. Lin and A.C. Estes (1997) Life- cycle cost design of deteriorating Structures. *Journal of Infrastructure Systems*, ASCE **123**(10), 1390–1401.
Frier, C. and J.D. Sorensen (2005) Stochastic simulation of chloride ingress into reinforced concrete structures by means of multi-dimensional Gaussian random field, *Proceedings of the 9th international conference on structural safety and reliability*, Rome, Italy, June 19–23. Millpress Science Publishers, Rotterdam [CD-ROM].
Ho, R.C.T. (2005) How can we capitalize on the concept of PPP. *Proceedings of One Day Conference on Public Private Partnerships – Opportunities and Challenges*, 22 February, Hong Kong, 72–74.
Hong, T.H. and M. Hastak (2007) Evaluation and determination of optimal MR&R strategies in concrete bridge decks. *Automation in Construction* **16**(2), 165–175.
Jiang, M., R.B. Corotis and J.H. Ellis (2000) Optimal life-cycle costing with partial observability. *Journal of Infrastructure Systems* **6**(2), 0056–0066.
Jiang, Y. and K.C. Sinha (1989) Bridge service life prediction model using the Markov chain. *Transport Res. Rec.* **1223**, 24–30.

Johannesson, B. *et al.* (2007) Multi-species ionic diffusion in concrete with account to interaction between ions in the pore solution and the cement hydrates. *Materials and Structures* **40**, 651–665.

Johnson B., T. Powell and C. Queiroz (1998) Economic analysis of bridge rehabilitation options considering life-cycle costs. *Transportation Research Record.* No. 1624, Washington, D. C. 8–15.

Junnila, S. and A. Horvath (2003) Life-cycle environmental effects of an office building. *Journal of Infrastructure Systems* **9**(4), 157–166.

Kayser, J.R. (1988) *The effects of corrosion on the reliability of steel girder bridges*, PhD thesis. University of Michigan, Ann Arbor, MI.

Kirpatrick, T.J., R.E. Weyers, C.M. Anderson-cook and M.M. Sprinkel (2002) Probabilistic model for the chloride-induced corrosion service life of bridge decks. *Cement and Concrete Research* **32**(12), 1943–1960.

Kleiner, Y. (2001) Schedule inspection and renewal of large infrastructure assets. *Journal of Infrastructure Systems* **7**(4), 136–143.

Kwan, J. (2005) Public private partnerships; Public private dialogue. *Proceedings of One Day Conference on Public Private Partnerships – Opportunities and Challenges*, 22 February, Hong Kong, 1–4.

Lee, K.M., H.N. Cho and C.J. Cha (2006) Life-cycle cost-effective optimum design of steel bridges considering environmental stressors. *Engineering Structures* **28**, 1252–1265.

Leung, W.C. and T.K. Lai (2002) Maintenance strategy of reinforced concrete structures in marine environment in Hong Kong, *Proceedings of International Congress – Challenge of concrete construction.* University of Dundee 5–11 Sept., 277–286.

Li, C.Q. (2003) Life-cycle modeling of corrosion-affected concrete structures: propagation. *Journal of Structural Engineering*, ASCE **129**(6), 753–761.

Li, C.Q., Y. Yang and R.E. Melchers (2008) Prediction of reinforcement corrosion in concrete and its effects on concrete cracking and strength reduction. *ACI Materials Journal* **105**(1), January–February, 3–10.

Liang, M.T., L.H. Lin and C.H. Liang (2002) Service life prediction of existing reinforced concrete bridges exposed to chloride environment. *Journal of Infrastructure Systems* **8**(3), 76–85.

Liu Y. and R.E. Weyers (1996) Time to cracking for chloride induced corrosion in reinforced concrete. In *Corrosion of Reinforcement in Concrete Construction*, Page, C. L. *et al.*eds. Royal Society of Chemistry, Cambridge, UK, 88–104.

Liu, M. and D.M. Frangopol (2006) Optimizing bridge network maintenance management under uncertainty with conflicting criteria life-cycle maintenance, failure and user costs. *Journal of Structural Engineering* **132**(11), 1835–1845.

Liu, Y. and R.E. Weyers (1998) Modeling the time-to-corrosion cracking in chloride contaminated reinforced concrete structures. *ACI Materials Journal* **9**(6), 675–681.

Lounis, Z. (2003) Probabilistic modeling of chloride contamination and corrosion of concrete bridge structures. *Proceedings of the 4th international symposium on uncertainty modeling and analysis (ISUMA' 03).*College Park, MD, p. 447.

Lounis, Z. and L. Daigle (2007) Environmental Benefits of Life Cycle Design of Concrete Bridges. *Proceedings of the 3rd International Conference on Life Cycle Management*, Zurich, Switzerland, Aug. 27–29, paper no. 293, pp. 1–6.

Lounis, Z., B. Martin-Perez and O. Hunaidi (2001) *Decision support tools for life prediction and rehabilitation of concrete bridge decks*. APWA International Public Works Congress, Philadelphia, PA, NRCC-4559.

Louniz, Z. and L. Amleh (2005) Reliability-based prediction of chloride ingress and reinforcement corrosion of aging concrete bridge decks. *ASCE Life Cycle Performance of Deteriorating Structures*, 113–122.

Ma, J., A. Chen and J. He (2009) General framework for bridge life cycle design. *Frontiers of Architecture and Civil Engineering in China* **3**(1), 50–56.

Macdougal, C., M.F. Green and S. Shillinglaw (2006) Fatigue damage of steel bridges due to dynamic vehicle loads. *Journal of Bridge Engineering*, ASCE **11**(3), 320–328.

Mak, C.K. and S. Mo (2005) Some aspects of the PPP approach to transport infrastructure development in Hong Kong. *Proceedings of One Day Conference on Public Private Partnerships – Opportunities and Challenges*, 22 February, Hong Kong.

Mangat, P. (1991) The effect of reinforcement corrosion on the performance of concrete structures, *BREU P3091 Repo*rt. University of Aberdeen, Aberdeen, Scotland.

Melchers, R.E., C.Q. Li, and W. Lawanwisut (2008) Probabilistic modeling of structural deterioration of reinforced concrete beams under saline environment corrosion. *Structural Safety* **30**, 447–460.

Miyamoto, A. and D.A. Frangopol (Ed) (2002) Maintaining the safety of deteriorating civil infrastructures. *Practical Maintenance Engineering Institute of Yamaguchi Univ*ersity, Yamaguchi, Japan.

Mohamadi, J., S.A. Guralnick and R. Polepeddi (1998) Bridge fatigue life estimation from field data. *Practical Periodical on Structural Design and Construction*, ASCE **3**(3), 128–133.

Mori, Y. and B.R. Ellingwood (1992) Reliability-based service-life assessment of aging concrete structures. *Journal of Structural Engineering* **119**(5), 1600–1621.

Moringa, S. (1988) Prediction of service lives of reinforced concrete buildings based on rate of corrosion of reinforcing steel. Shimizu Corp., Tokyo, Report no. 23.

Ng, S.T., Y.M.W. Wong and M.M. Kumaraswamy (2005) Experience of a PPP-based high speed road maintenance project in Hong Kong. *Systematic innovation in the management of construction projects and processes*, Kazi A. S. ed. Technical Research Centre of Finland and Association of Finnish Civil Engineers, 251–262.

Novick, D. (1990) Lifecycle considerations in urban infrastructure engineering. *Journal of Management in Engineering*, ASCE **6**(2), 186–196.

Nowak A.S. (1993) *Calibration of LRFD bridge design code*. Final Report prepared for NCHRP, TRB, National Research Council, Dept. of Civil Engineering, University of Michigan, Ann Arbor, MI.

Obata, T., A. Nakajima and K. Maeda (2007) Study on application of reliability analysis for bridge management system on steel bridge structures. *Life-Cycle Cost and Performance of Civil Infrastructure Systems*, Cho, Frangopol and Ang. eds., Taylor & Francis Group, London, UK, p. 291.

OECD (1983) *Bridge rehabilitation and strengthening*. Organization for Economic Co-operation and Development, Road Transportation Research Report, Paris, France.

Papadakis, V.G. *et al*. (1996) Mathematical modeling of chloride effect on concrete durability and protection measures. In: *Concrete repair, rehabilitation and protection*, R.K. Dhir and M.R. Jones, eds. E & FN Spon, London, UK, 165–174.

Queensland (2005) Another $2 billion for Queensland infrastructure. *Engineers Australia*, **77**(11), p. 25.

Radomski, W. (2002) *Bridge Rehabilitation*. Imperial College Press, London, UK.

Rebitzer, G., D. Hunkeler and O. Jolliet (2003) LCC – the economic pillar of sustainability: methodology and application to wastewater treatment. ENAC – Life Cycle System, Swiss Federal Institute of Technology Lausanne, Lausanne, Switzerland.

Reigle, J.A. and J.P. Zaniewski (2005) Risk-based life-cycle cost analysis for project-level pavement management. *Transportation Research Record* **1816**, 34–42.

Roberts, M.B. and C. Middleton (2000) A proposed empirical corrosion model for reinforced concrete. *Proc. Instn Civ. Engrs Structs & Bldgs* **140**(1), February, 1–11.

Roelfstra, G., R. Hajdin, B. Adey and E. Bruhwiler (2004) Condition evolution in bridge management systems and corrosion-induced deterioration. *Journal of Bridge Engineering* **9**(3), 1 May, 268–277.

Salem, O., S. AbouRizk and S. Ariaratnam (2003) Risk-based life-cycle costing of infrastructure rehabilitation and construction alternatives. *Journal of Infrastructure Systems* **9**(1), 6–15.

Saloranta, A. (1993) High performance coatings for steel bridges. *Progress in Organic Coating* **22**, 345–355.

Sarja, A. and E. Vesikari (1996) Durability design of concrete structures. *Report of the Technical Committee 130-CSL*, RILEM.

Setunge, S., A. Kumar, A. Nezamian, S. De Sliva and W. Lokuge (RMIT); Carse, A.; Spathonis, J. and Chandler, L. (QDMR); Gilbert, D. (QDPW); Johnson, B. (Ove Arup); Jeary, A. (UWS); and L. Pham (CSIRO) (2005) Whole of life cycle cost analysis in bridge rehabilitation. *Report 2002-005-C-03*. RMIT University, Melbourne, Australia.

Shepard, R. and I. Abed-Al-Rahim (1993) Using life-cycle cost analysis for bridge management system. *FHWA Life Cycle Cost Sym.*, Federal Highway Administration, Washington, D. C.

Siemes, A., A. Vrouwenvelder and A. Van der Beukel (1985) Durability of buildings: a reliability analysis. *Heron* **30** (3), 3–48.

So, K.K.L., M.M.S. Cheung and E.X.Q. Zhang (2009) Life-cycle cost management of concrete bridges. *Proceedings of the Institution of Civil Engineers – Bridge Engineering* **162**, September, Issue BE3, 103–117

Sommer, A.M., A.S. Nowak and P. Thoft-Christensen (1993) Probability-based bridge inspection strategy. *Journal of Structural Engineering,* ASCE **119**(12), 3520–3536.

Stewart, M.G. (2001) Reliability-based assessment of ageing bridges using risk ranking and life cycle cost decision analyses. *Reliability Engineering and System Safety* **74**, 263–273.

Stewart, M.G. (2004) Spatial variability of pitting corrosion and its influence on structural fragility and reliability of RC beams in flexure. *Structural Safety* **26**(4), 453–470.

Stewart, M.G. and J.A. Mullard (2007) Spatial time dependent reliability analysis of corrosion damage and the timing of first repair for RC structures. *Engineering Structures* **29**, 1457–1464.

Stewart, M.G. and D.V. Rosowsky (1998) Time-dependent reliability of deteriorating reinforced concrete bridge decks. *Structural Safety* **20**(1), 91–109.

Sudret, B. (2008) Probabilistic models for the extent of damage in degrading reinforced concrete structures. *Reliability Engineering and System Safety* **93**, 410–422.

Suo, Q. and M.G. Stewart (2009) Corrosion cracking prediction updating of deteriorating RC structures using inspection information. *Reliability Engineering and System Safety* **94**, 1340–1348.

Suwito, C. and Y. Xi (2006) The effect of chloride-induced steel corrosion on service life of reinforced concrete structures. *Structure and Infrastructure Engineering* **4**(3), 177–192.

Tam, C.K. and S.F. Stiemer (1996) Development of bridge corrosion cost model for coating maintenance. *Journal of Performance of Constructed Facilities*, May, 47–56.

Tavakkolizadeh, M. and H. Saadatmanesh (2003) Fatigue strength of steel girders strengthened with carbon fiber reinforced polymer patch. *Journal of Structural Engineering* **129**(2), 186–196.

Thoft-Christensen, P. and J.D. Sorensen (1987) Optimal strategy for inspection and repair of structural systems. *Civil Engineering System* **4** (June), 94–100.

Tighe, S. (2001) Guidelines for probabilistic pavement life cycle cost analysis. *Transportation Research Record* **1769**, 28–38.

Tuutti, K. (1982) Corrosion of steel in concrete. *CBI Research Report* **4**(82), Swedish Cement and Concrete Research, Institute, Stockholm, Sweden.

Ugwu, O.O., M.M. Kumaraswamy, F. Kung and S.T. Ng (2005) Object-oriented framework for durability assessment and life cycle costing of highway bridges. *Automation in Construction* **14**, 611–632.

Val, D.V. (2005) Effect of different limit states on life-cycle cost of RC structures in corrosion environment. *Journal of Infrastructure System*, ASCE **11**(4), December 1.

Val, D.V. and M.G. Stewart (2003) Life-cycle cost analysis of reinforced concrete structures in marine environments. *Structural Safety* **25**(4), 343–362.

Val, D.V. and P.A. Trapper (2008) Probabilistic evaluation of initiation time of chloride-induced corrosion. *Reliability Engineering and System Safety* **93**, 364–372.

Vesikari, E. (1988) *Service Life of Concrete Structures with regard to Corrosion of Reinforcement*. Technical Research Centre of Finland, ESPOO 1988.

Vu, K.A.T. and M.G. Stewart (2000) Structural reliability of concrete bridges including improved chloride-induced corrosion model. *Structural Safety* **22**(4), 313–333.

Vu, K.A.T. and M.G. Stewart (2002) Spatial variability of structural deterioration and service life prediction of reinforced concrete bridges. *Proceedings of 1st International Conference on Bridge Maintenance, Safety and Management (IABMAS'02)*, Barcelona, Spain (CD-ROM). [No page range is listed]

Vu, K.A.T., M.G. Stewart and J.A. Mullard (2005) Corrosion induced cracking: experimental data and predictive models. *ACI Structural Journal* **102**(5), 719–726

Wang, E. (2005) Infrastructure rehabilitation management applying life-cycle cost analysis. *Proceedings of the 2005 ASCE International Conference on Computing in Civil Engineering*.

Weyers, R.E. (1998) Service life model for concrete structures in chloride-laden environments. *Journal of ACI Materials* **95**(5), 546–557.

Weyers, R.E., B. Prowell, M.M. Sprinkel and M. Vorster (1993) *Concrete bridge protection, repair, and rehabilitation relative to reinforcement corrosion: a methods application manual*. Strategic Highway Research Program, Washington, DC, SHRP-S-360.

Williamson, G.S., R.E. Weyers, M.M. Sprinkel and M.C. Brown (2008) Concrete and steel type influence on probabilistic corrosion service life. *ACI Materials Journal* **106**(1), January–February, 82–88.

Wirahadikusumah, R., D. Abraham and T. Iseley (2001) Challenging issues in modeling deterioration of combined sewers. *Journal of Infrastructure Systems* **7**(2), 77–84.

Yang, S.I., D.M. Frangopol and L.C. Neves (2006) Optimum maintenance strategy for deteriorating bridge structures based on lifetime functions. *Engineering Structures* **28**, 196–206.

Yescombe, E.R. (2007) *Public-Private Partnerships – Principles of Policy and Finance*, Elsevier Limited.

Zhang, J. and Z. Lounis (2006) Sensitivity analysis of simplified diffusion-based corrosion initiation model of concrete structures exposed to chlorides. *Cement and Concrete Research* **36**, 1312–1323.

Zhang, X.Q. (2004) Concessionaire selection: methods and criteria. *Journal of Construction Engineering and Management*, ASCE **130**(2), 235–244.

Zhang, X.Q. and S.M. AbouRizk (2006) Determining a reasonable concession period for private sector provision of public works and service. *Canadian Journal of Civil Engineering* **33**(5), 622–631.

Zhang, X.Q. and M.M. Kumaraswamy (2001) Procurement Protocol for Public Private Partnered Project. *Journal of Construction Engineering and Management* **127**(5), 137–142.

Zhao, Z., A. Haldar and F.L. Breen Jr (1994) Fatigue-reliability evaluation of steel bridges. *Journal of Structural Engineering* **120**(5), May, 1608–1623.

Chapter 6
Life cycle analysis of highway composite bridges

H. Gervásio

6.1 Introduction

Highway bridges are important assets in the infrastructure network, and the cost of construction and operation of highway bridges are key issues for bridge engineers. However, in light of sustainable construction, the design of bridges requires more than complying with safety requirements and economical constraints. The construction, operation and ultimately the way the bridge is demolished have a significant impact on the environment, in the surrounding population and in the users of the bridge. Therefore, the multidimensional perspective of sustainability requires the combination of current design criteria with other important aspects such as society and environment, preferably considering a life cycle analysis (LCA).

LCA usually refers to the quantification of potential environmental impacts of a product or a system throughout its life cycle (ISO 14040, 2006). LCA was initially used for the assessment of simple products, usually with short life spans, and therefore the methodology was not specific for construction assets. However, over the last years, and in the perspective of sustainable construction, the approach has been adapted and successfully applied to construction systems (Gervásio and Simões da Silva, 2013a).

On the other hand, life cycle cost analysis (LCCA) and whole life costing (WLC), are approaches much more familiar to the construction sector, although the two are often confused. According to ISO 15686-5 (2008), LCC includes the cost of planning, design, acquisition, operation, maintenance and disposal of buildings and other construction assets, while WLC includes, apart from those, incomes and other costs such as non-construction costs and externalities. Hence, LCC may be considered as a subset of WLC.

It is observed that in WLC, social, environmental or business costs or benefits are considered as externalities. This could lead to a double-counting of impacts when WLC is used together with LCA. Hence, hereafter only LCC is considered.

The work herein presented was developed with the aim of integrating in an LCA, environmental, economical and social criteria, following the guidance provided by the standardized framework for LCA (ISO 14040, 2006 and ISO 14044, 2006). The developed approach focuses on the assessment of highway bridges.

Finally, the role of bridge engineering in sustainable construction is illustrated by a case study: a composite bridge crossing a highway.

6.2 Integrated life cycle analysis of bridges

6.2.1 General framework

CEN TC350 *Sustainability of construction works* was mandated for the development of standardized methodologies for the assessment of sustainability issues of new and existing buildings. As a result, in recent years a series of voluntary horizontal standards have been published. Contrary to the life cycle assessment of buildings, there is not a standardized methodology for LCA of bridges, although CEN TC350 has recently extended its mandate to include the assessment of other civil engineering works.

The life cycle environmental analysis (LCEA) has currently the most well established standardized framework, although there is still no generalized accepted methodology in the scientific community. In a decreasing order of development follows the life cycle economical analysis and the life cycle social analysis (LCSA).

For those reasons, the integral life cycle approach herein presented is based on the standardized framework for LCEA, according to the ISO standards 14040 (2006) and 14044 (2006), although extended in order to integrate LCCA and LCSA (Gervásio, 2010).

Therefore, the generalized approach entails the four main steps of the ISO standard 14040 (2006): (1) the goal and scope step; (2) the inventory step; (3) the impact assessment step; and (4) the interpretation step (see Figure 6.1). However, each step of the analysis is adapted in order to allow the integration of economical and social aspects in the LCA.

The integration of the three criteria is based on the condition that the evaluation of the three criteria share the same goal and scope and they are based on the same inventory analysis (Gervásio, 2010). Furthermore, the application of a multicriteria approach enables to combine the three criteria before the interpretation step; however, this aspect is not further addressed in this text.

Hence, the work presented focuses on the quantification of the three criteria, as indicated by the dashed box in Figure 6.1.

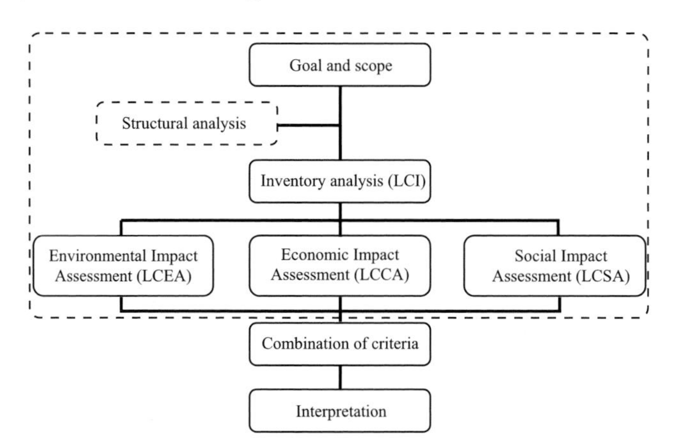

Figure 6.1 *Flowchart of the life cycle integral analysis (Courtesy of Gervásio and Simões da Silva, 2013a).*

6.2.2 Goal and scope of the integrated analysis

The goal of the integrated approach is to enable the identification of the most critical stage(s) and process(es) throughout the bridge life cycle, in relation to the criteria considered, so that adequate measures may be considered in order to enhance its lifetime performance.

The LCA covers all stages from raw material acquisition to decommission and deposition of demolition waste, through construction and operation of the bridge (Figure 6.2). Each box in Figure 6.2 represents a unit process. In fact, each box involves a number of subunit processes (e.g. the production of cement for input to the production of concrete); however, for simplification, the boxes presented are assumed to be the main unit processes.

Moreover, it is assumed that the bridge is designed according to the corresponding Eurocode. In this case, EN1990 (2002) specifies a service life of 100 years, which may be considered the time span for the LCA.

Additionally, in this stage, it is necessary to define the functional equivalent, i.e. the reference unit in the analysis to which the input and output data are normalized (ISO 14044, 2006). The functional equivalent of a bridge is a representation of its required technical characteristics and functionalities. This unit is also the basis for comparative LCA.

6.2.3 Lifetime structural analysis

Over the time span of the analysis, the structural behaviour of the bridge deteriorates and maintenance actions are needed to keep the bridge above the required condition throughout its service life (Gervásio and Simões da Silva, 2013a). Therefore, in an LCA, it is necessary to estimate the service life of the bridge and bridge components in order to anticipate the frequency of the required actions and respective costs.

Although the prediction of the lifetime behaviour of the bridge may be based on more or less complex numerical modelling (Gervásio and Simões da Silva, 2013a), in this methodology, a scenario-based approach is considered, which enables to take into account different maintenance types and frequencies over the service life of the structure. This issue will be further elaborated in the case study.

6.2.4 Inventory analysis

The inventory analysis entails the quantification of the inventory flows for a product system, in this case, a bridge. In traditional LCEA the inventory flows include inputs of water, energy and raw materials, and releases to air, land and water. Taking into consideration the aim of the approach, economical and social data are also considered in the inventory analysis. This implies that input and/or output flows relevant for the quantification of these criteria need also to be collected. A summary of inventory flows for each unit process is represented in Figure 6.3.

An important aspect that should be addressed by life cycle inventory analysis is the allocation of input/output flows of materials with recycling content. This issue is

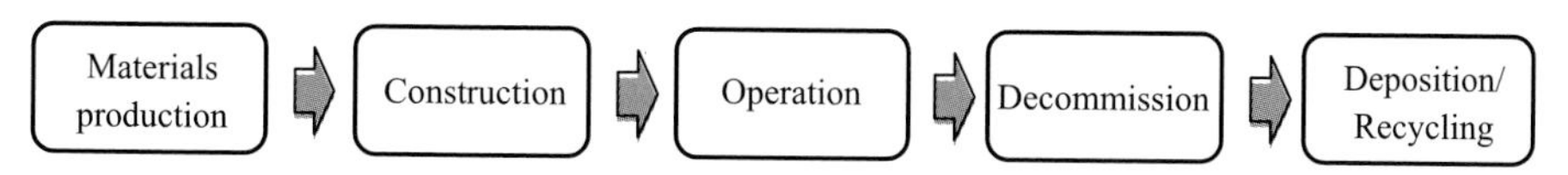

Figure 6.2 *System boundary of the integral analysis.*

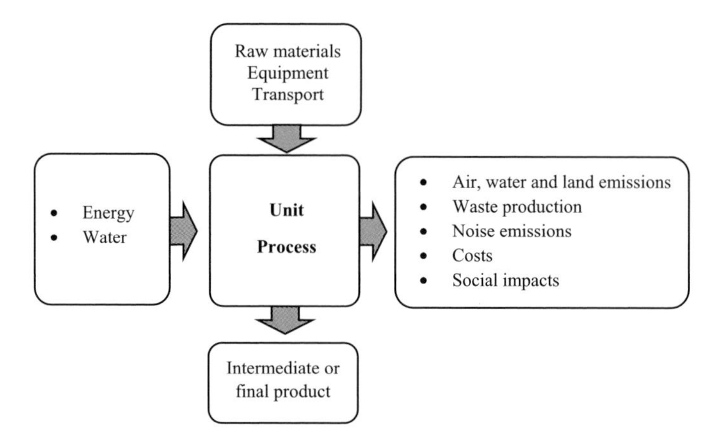

Figure 6.3 *Inventory data of a general unit process.*

particularly important in a construction system where the reuse and recycling of construction materials are often acknowledged as contributing to the sustainability of the sector. This aspect will be further detailed in the case study.

6.2.5 Impact assessment

In the impact assessment stage, the inventory data collected in the previous stage is processed according to the selected indicators for the environmental, economical and social analyses.

The set of indicators for each analysis were chosen according to the goals of the integral approach. The indicators adopted for the environmental, economical and social analysis are represented in Table 6.1.

However, in this chapter, the criteria of "aesthetics" and "noise emissions" are not further addressed since they are not used in the case study; therefore, for the quantification of impacts on society only user costs are hereafter considered.

6.2.5.1 Quantification of environmental indicators

This stage aims for the evaluation of the significance of potential environmental impacts using the results of the life cycle inventory analysis. In general, this process involves associating inventory data with specific environmental impacts, and is made of two parts: (1) classification and (2) characterization.

The classification implies a previous selection of appropriate impact categories, according to the goal of the study, and the assignment of the inventory flows to the selected impact categories. Characterization factors are then used representing the relative contribution of an inventory flow to the impact category indicator. According to this method impact categories are linear functions, i.e. characterization factors are independent of the magnitude of the environmental intervention, as given by expression 6.1:

$$impact_{cat} = \sum_{i} m_i \times charact\_factor_{cat,i} \qquad (6.1)$$

where m_i is the mass of the inventory flow i and $charact\_factor_{cat,\,i}$ is the characterization factor of inventory flow i for the impact category.

Table 6.1 *Indicators for environmental and economical assessment.*

Environmental	*Economical*	*Social*
Abiotic depletion	Initial costs	User costs
Acidification	Operation, maintenance	Aesthetics
Eutrophication	and repair costs	Noise emissions
Global warming	End-of-life costs	
Ozone layer depletion		
Photochemical oxidation		

Data source: Gervásio and Simões da Silva (2013a).

In this case, six environmental categories were selected, as indicated in Table 6.1: abiotic depletion (in kg Sb eq.), global warming (in kg CO_2 eq.), ozone layer depletion (in kg CFC-11 eq.), photochemical oxidation (in kg C_2H_4), acidification (in kg SO_2 eq.) and eutrophication (in kg PO_4- eq.). Then, the quantification of the selected environmental categories is made using the characterization factors of the CML methodology (Guinée *et al.*, 2001).

6.2.5.2 Quantification of economic criteria

The indicators selected for the economical aspect aim to represent the direct costs incurred by the highway administration over the study period; namely, the construction cost, the cost of maintenance and repairs throughout the service life and the cost of demolition and disposal. These costs, hereafter denominated direct costs, represent the period of interest that the cost analysis is aimed at.

These costs occur in different time periods, over the life cycle of the bridge. Therefore, future costs occurring over the life cycle of the bridge are discounted to their present-value, as of the base year, by using expression 6.2 (Fuller and Petersen, 1996):

$$NPC = \sum_{t=0}^{N} \frac{C_t}{(1+d)^t} \tag{6.2}$$

where *NPC* is the net present cost; C_t is the sum of all relevant costs, less any positive cash flows, occurring in year t; N is the number of years in the study period; and d is the discount rate.

6.2.5.3 Quantification of user costs

The social aspect of the proposed approach aims to model user costs that are directly related to the bridge. The quantification of user costs due to traffic flow that passes under and above the bridge in normal traffic conditions does not capture the interaction between the bridge and its users. This interaction is captured when the bridge is subjected to any work activity and traffic congestion results from delays in the construction zone. Thus, user costs are the additional costs resulting from construction work being performed on the bridge, which may affect the traffic over the bridge, under the bridge or both. Then, three types of user costs are hereafter considered for all stages: the cost of vehicle operation, the cost of driver delay and the cost of safety. User costs are quantified taking into account the hourly traffic in the motorway, the characteristics of the traffic and the characteristics of the work zone by the use of the QUEWZ98 model (Seshadri and Harrison, 1993).

Driver's delay cost (DDC) is the cost of the time lost by a driver while travelling through a work zone. This cost is given by the difference between the cost of the time lost by a driver while travelling at normal speed and the time lost while travelling at a reduced speed due to construction works on the same length of the motorway, as given by expression 6.3 (Gervásio, 2010):

$$DDC = \left(\frac{L}{S_a} - \frac{L}{S_n} \right) \times ADT \times N \times \sum_i \left(DTC_i \times p_i \right) \qquad (6.3)$$

where L is the length of affected motorway (in km), S_a is the traffic speed during work activity (km/h), S_n is the normal traffic speed (km/h), ADT is the average daily traffic (cars/day), N is the number of days of road work, DTC_i is the cost per hour of a driver's time of a class i vehicle and p_i is the percentage of class i vehicles in total traffic flow.

Construction related delays result in additional costs also for the owner of the vehicle. These additional costs are denominated vehicle operating costs (VOC). This cost is given by the difference between the cost of the operation of the vehicle while travelling at normal speed and the operation of the vehicle while travelling at a reduced speed due to construction works on the same length of the motorway, as given by expression 6.4 (Gervásio, 2010):

$$VOC = \left(L - \frac{S_a}{S_n} \times L \right) \times ADT \times N \times \sum_i \left(VOC_i \times p_i \right) \qquad (6.4)$$

where VOC_i is the operation cost of class i vehicle (€/km) and the remaining parameters have the same meaning as expression 6.3.

Finally, accident costs (AC) represent the costs of additional accidents due to a work zone in the motorway; thus, they are calculated by the difference between the cost of accidents in a length of motorway with no work activity and the cost of accidents in the same length when there is work activity, as given by the following expression,

$$AC = L \times ADT \times N \times \left(R_a - R_n \right) \times C_a \qquad (6.5)$$

where R_a and R_n are the accident rates (per vehicle-kilometre) during work activity and under normal traffic conditions, respectively; C_a is the cost per accident; and the remaining parameters have the same meaning as expression 6.3.

Likewise, user costs occur in different time periods over the life cycle of the bridge. Therefore, future costs are discounted to their present-value by using expression 6.2.

6.3 Life cycle analysis of a motorway bridge

The aim of this section is to illustrate the application of the integrated life cycle approach to a highway composite bridge. The case study is split into two parts. In the first part, the original case study is analysed and the most critical stages and processes are identified taking into account the selected criteria. Then, an alternative case study is presented aiming to improve the lifetime performance of the bridge.

6.3.1 Scope and functional equivalent of case study

The scope of the LCA performed is illustrated in Figure 6.2 and each box will be considered next.

The time period of analysis coincides with the design service life of the bridge; thus, a time span of 100 years is considered. In addition, it is assumed that material production and construction of the bridge occur in the first year and that the demolition and deposition of the structure will occur in the 100th year. The functional equivalent adopted in this case study is a bridge overcrossing a motorway, with a capacity up to four lanes of traffic in each direction, designed for a service life of 100 years.

6.3.2 Description of original case study

6.3.2.1 Project details

The object of this case study is an overpass built over the dual carriageway A1—*Auto Estrada do Norte*, in Portugal (Figure 6.4). Each carriageway has a width of 16.30 m and it comprises a main running surface with 11.25 m, a hard shoulder along the inner lane with a width of 4.05 m and a hard shoulder along the outer lane with 1.00 m. When the bridge is opened to traffic there are 3 lanes of traffic, with a width of 3.75 m, in each carriageway; although, it may be widened to 4 lanes. The central reservation is 4.00 m wide.

The bridge is made of a composite steel-concrete structure with three spans of 18.50 m, 40.80 m and 18.50 m, as illustrated in Figure 6.5. The deck is fully restrained against displacements over the middle piers and simply supported in the abutments.

The middle piers are made of cast in-situ concrete, with a circular cross-section with diameter of 1.20 m. The cross-section of the bridge is made of two steel girders with a total height of 1.35 m. The girders are braced every 5.2 m in the main span and every 5 m in the side spans.

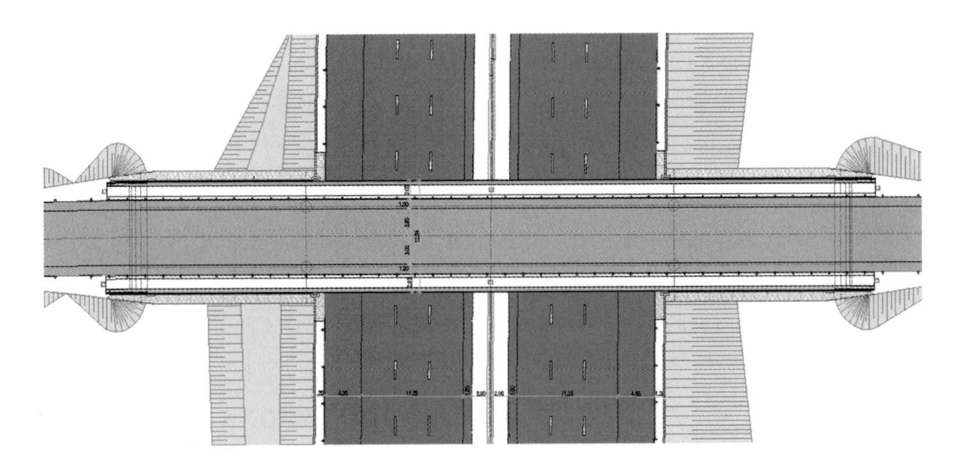

Figure 6.4 *Plan view of the motorway bridge.*

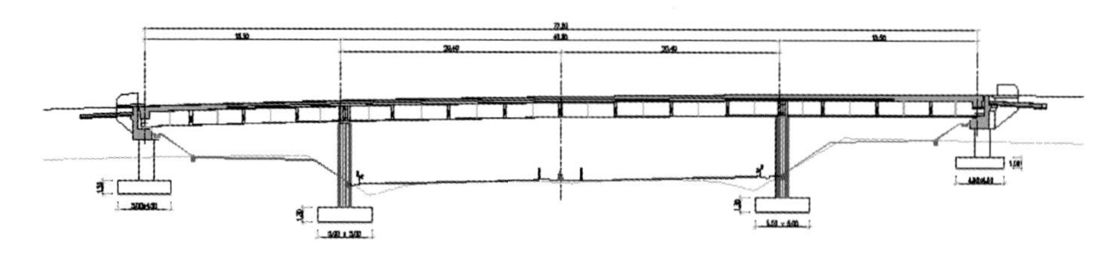

Figure 6.5 *Elevation view of the motorway bridge.*

The transversal profile of the deck comprehends two lanes of traffic, one in each direction, with a total width of 12.04 m. The total area of the deck is 936.71 m².

The bill of the main materials of the bridge and respective unit costs are represented in Table 6.2.

6.3.2.2 Construction details

The construction of the bridge took a total of 87 days, from 3-08-07 to 28-10-07, according to the planning represented in Table 6.3, which was provided by the Portuguese highway concessionaire BRISA.

The abutments and piers are cast in-situ to enable the installation of the steel structure. The steel structure is assembled in two half-parts, next to the motorway. Each part of the steel structure is then installed in its final position by the aid of cranes. After

Table 6.2 *Bill of main materials and costs.*

Component	*Material*	*Quant.*	*Cost*
Foundation regularization	Concrete grade C16/20	29.00 m³	49.32 €/m³
Foundation bases	Concrete grade C25/30	223.00 m³	58.31 €/m³
Abutments elevation	Concrete grade C30/37	123.00 m³	63.19 €/m³
Piers	Concrete grade C30/37	33.00 m³	63.19 €/m³
Deck Precast slab	Concrete grade C35/45	161.00	58.04 €/m³
	Reinforcement steel grade A500	58.31 ton	0.65 €/kg
Deck Cast "in situ" slab	Concrete grade C35/45	116.00 m³	70.81 €/m³
	Concrete grade C35/45 (low shrinkage)	48.00 m³	93.69 €/m³
	Reinforcement steel grade A500	17.86 ton	0.65 €/kg
Girders	Steel grade S355	145.68 ton	1.19 €/kg
Shear studs	Steel grade F_{uk}=350 MPa	3328 un	0.73 €/un
Coating of steel	Paint	1,296 m²	10.80 €/m²
Side-walks	Light-weight concrete	38.00	63.80 €/m³
All elements except deck	Reinforced steel grade A500	57.60 ton	0.65 €/kg

Table 6.3 *Plan of construction.*

Main activity	*Initial date*	*Final date*
Excavation	03-08-07	22-10-07
Casting of abutments	09-08-07	28-10-07
Piers	09-08-07	25-08-07
Foundation bases	09-08-07	20-08-07
Piers	14-08-07	25-08-07
Installation of supports	26-08-07	01-09-07
Assembling of steel structure	30-08-07	13-09-07
Preassembling of steel structure in the ground	30-08-07	04-09-07
Erection of work form in middle span	01-09-07	04-09-07
Erection of steel structure	05-09-07	06-09-07
Welding middle span	07-09-07	08-09-07
Coating middle span	09-09-07	11-09-07
Removal of work form	12-09-07	13-09-07
Construction of concrete deck	14-09-07	10-10-07
Erection of slab panels	14-09-07	17-09-07
Settlement of steel structure	18-09-07	18-09-07
Sealing of concrete panels	19-09-07	22-09-07
Casting "in situ" of deck (near abutments)	23-09-07	04-10-07
Casting of remaining parts of deck	05-10-07	10-10-07
Finishing	11-10-07	28-10-07

the installation of the two parts of the structure, the longitudinal girders are welded in the middle of the central span. Afterwards the full-deck-width precast concrete panels are installed. These precast concrete panels are connected to the steel girders by studs located in pockets in the panels. The panels are joined transversally by overlapping hoop bars that project from each edge of the panels. Individual bars are threaded within the loop bars to complete the connections, which are then encased in concrete. These joints are sealed with a low shrinkage concrete that is cast "in situ".

The concrete slab near the abutments, in each side of the structure, is also cast "in situ". Finally all the remaining deck equipment is installed in the bridge.

6.3.2.3 Traffic constraints during construction works

During the construction of the bridge and according to information provided by the contractor, the motorway was always in service, although the traffic flow was subjected to various constraints. Being a new bridge, there was no traffic above the bridge.

According to Portuguese legislation, traffic conditions through work zones in major motorways are subjected to several restrictions, which include that (Gervásio, 2010).

1. there should always be two lanes of traffic in each direction;
2. the width of the carriageway, in the work zone, cannot be less than two thirds of the original width;
3. the maximum speed through the work zone must be higher than two thirds of the speed under normal conditions in the same location.

It is assumed that during the construction of the bridge, these restrictions were fully taken into account. Therefore, there were always three lanes of traffic in each direction of the motorway, except during the erection of the composite deck (6 days) (see Table 6.3). In this case, it was necessary to cancel the traffic in one carriageway and divert the traffic to the other one, resulting in one lane of traffic in each direction. Moreover, during most of the period of construction it was considered that the traffic in both carriageways was constrained to a maximum speed of 80 km/h.

6.3.3 Scenarios and assumptions for life cycle analysis

6.3.3.1 Traffic under and over the bridge

An average daily traffic (ADT) of 48,862 vehicles per day was assumed in the base year of the study. It was considered that the percentages of lightweight vehicles and

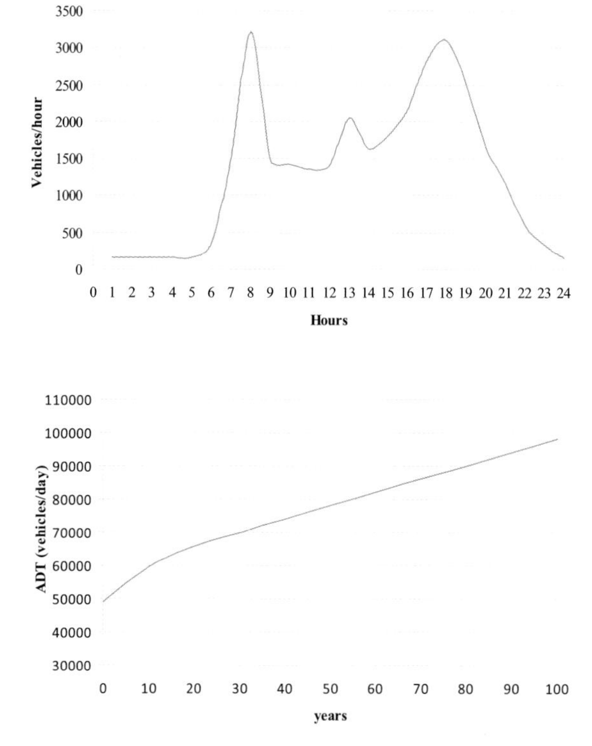

Figure 6.6 *Distribution of hourly traffic: (a) hourly traffic distribution; (b) traffic growth over the time.*

heavyweight vehicles were 88% and 12% of the ADT, respectively. Moreover, it was also considered that the traffic during the night period (2200–0700) is about 10% for lightweight vehicles and 8% for heavyweight vehicles. A distribution of hourly traffic was assumed for the motorway, which is represented in Figure 6.6a.

In addition, it was necessary to estimate traffic growth over the period of time of the analysis. In this case, the following scheme was considered for the traffic growth over time (see Figure 6b): a growth rate of 2% until year 10, a rate of 1% until year 20 and a rate of 0.5% after year 20. According to this trend, approximately in 10 years from the base year, the traffic growth reaches 60,000 vehicles per day, which means that the highway, according to Portuguese requirements, will have to be widened to

Table 6.4 *Timetable of inspection activities of the bridge.*

Type of inspection	*Frequency*	*Cost*
Routine inspections	annually	0.86 €/m²
Principal inspections	Every 6 years	8.7 €/m²
Special inspections	2 in 100 years	23.0 €/m²

Data source: SBRI (2013).

Table 6.5 *Timetable of maintenance/rehabilitation activities of the bridge.*

Damage	*Maintenance action*	*Years of maintenance action*	*Cost*
Steel girders			
Corrosion (small points/ small areas)	Partial replacement	25, 60, 95	55.00 €/m²
Corrosion (total surface)	Complete renewal	35, 70	75.00 €/m²
Concrete deck			
Corrosion of the reinforcement deck	Partial renewal	25, 50, 75	350.00 €/m²
Concrete edge beam major defects	Total replacement	40, 80	355.00 €/m
Concrete edge beam minor defects	Partial renewal	25, 65	100.00 €/m
Expansion joints			
Broken modules (considering a modular joint)	Total replacement	40, 80	1,580.00 €/m
Broken concrete header	Total/partial replacement	10, 20, 30, 50, 60, 70, 90	80.00 €/m

(*Continued*)

Table 6.5 *Continued*

Damage	*Maintenance action*	*Years of maintenance action*	*Cost*
Dust	Cleaning	10, 20, 30, 50, 60, 70, 90	1.50 €/m
Bearings			
Used up	Total replacement	35, 70	1,430.00 €/un
Minor defects	Partial replacement	20, 55, 90	350.00 €/un
Corrosion of metallic elements	Painting of metallic elements	35, 70	465.00 €/un
Road surface			
Cracks, ruts, excavation (total area)	Total replacement	20, 40, 60, 80	11.50 €/m^2
Cracks, ruts, excavation (partial area)	Minor repairs	10, 30, 50, 70, 90	11.40 €/m^2
Water proofing layer			
Cracks, ruts, excavation	Total replacement	40, 80	40.00 €/m^2
Railings			
Used up	Total replacement	40, 80	140.00 €/m
Corrosion	Painting of metallic elements	20, 60	19.00 €/m
Safety barrier			
Used up	Total replacement	40, 80	420.00 €/m
Small defects	Minor repairs	25, 65	50.00 €/m^2

Data source: SBRI (2013).

four lanes of traffic in each carriageway. However, the widening of the motorway has no effect on the bridge, as the main span of the bridge is long enough to accommodate this widening.

6.3.3.2 Operation stage—maintenance and rehabilitation plan

This stage starts when the bridge comes into service and ends when it reaches the end of its service life. To enable LCA, and in the absence of more complex structural degradation models, scenarios are needed for the service life of bridge components, maintenance activities and respective costs.

Moreover, during the service life of a bridge two main actions are taken to ensure adequate safety: inspection and maintenance. Hence, in this case study, a plan for inspection and a plan for maintenance of composite bridges, with respective unit costs, were adopted and the corresponding timetables are illustrated in tables 6.4 and 6.5, respectively.

The adopted schemes were established in the framework of the European Research Project SBRI – *Sustainable Steel-Composite Bridges in Built Environment* (SBRI, 2013) and aims to represent an average bridge inspection and maintenance schemes within the participant countries (Portugal, Germany, France and Denmark). It is observed that these plans reflect the experience and expertise of the partners involved in the project, which included highway administrations, bridge constructors, bridge designers, consultants and academic researchers.

According to the time of intervention in the bridge for each activity, and in order to minimize the duration of work activity over and under the bridge, the activities happening in the same year are assumed to take place at the same time. The duration of each activity depends on the type of work to be done, the ability of the workforce and the capacity rate of construction equipment. Thus, for each activity (indicated in Table 6.5) a planning of the respective work was made based on the estimated bill of materials and on the estimated capacity rate for each activity. Based upon this information a period of time was estimated for each activity.

Whenever there's any activity on the bridge, traffic conditions over and below the bridge are affected and traffic restrictions in work zones are imposed as described before. Table 6.6 indicates the effect on traffic over and below the bridge.

Table 6.6 *Total duration (in days) of maintenance/rehabilitation activities.*

Year of activity	*Traffic restrictions*	
	Under the bridge	*Over the bridge*
10	No restrictions	1 lane closed / per day
20	No restrictions	1 lane closed / per day
25	1 lane closed / per day	Speed reduction
30	No restrictions	1 lane closed / per day
35	1 lane closed / per day	Speed reduction
40	1 lane closed / per day	1 lane closed / per day
50	1 lane closed / per day	1 lane closed / per day
55	No restrictions	Speed reduction
60	1 lane closed / per day	1 lane closed / per day
65	1 lane closed / per day	Speed reduction
70	1 lane closed / per day	1 lane closed / per day
75	1 lane closed / per day	Speed reduction
80	1 lane closed / per day	1 lane closed / per day
90	No restrictions	1 lane closed / per day
95	1 lane closed / per day	No restrictions

Table 6.7 *Duration (in days) of bridge demolition.*

Demolition stages	*Description*	*Duration (in days)*
1st stage	Preparation of base supports for scaffolding; Preparation of terrain for deposit of concrete pieces; Erection of scaffolding; Transversal cuts in deck; Longitudinal cuts of deck (partially).	10
2nd stage	Dismantling of one part of composite deck and deposition of the resulting pieces in the side terrain; Dismantling of other part of composite deck and deposition of the resulting pieces in the side terrain.	6
3rd stage	Dismantling of scaffolding in central span; Demolition of the concrete pieces from the deck; Demolition "in situ" of piers and abutments; Removal and transportation of materials; Cleaning of the area.	7

6.3.3.3 End-of-life plan of the bridge

At the end-of-life stage it was assumed that the bridge is demolished. It was estimated that a total of 23 days were needed (Table 6.7).

During the demolition of the bridge, it was considered that traffic over the bridge was diverted to an alternative road. Traffic under the bridge was conditioned by the demolition of the bridge; however, it was assumed that, except for the 2nd stage (see Table 6.7), there were always three lanes of traffic in each direction. During the second stage, traffic in the motorway would have to be diverted to only one carriageway as the carriageway below the part of the deck being dismantled had to be closed due to users' safety reasons.

In addition, after the demolition of the bridge and sorting of materials, the materials are sent to different places according to its characteristics and potential for recycling, as described further ahead in the text.

6.3.4 Life-cycle environmental analysis

Life cycle analysis (LCA) was performed according to the method described in Section 6.2.3.1 and using the software SimaPro (2008).

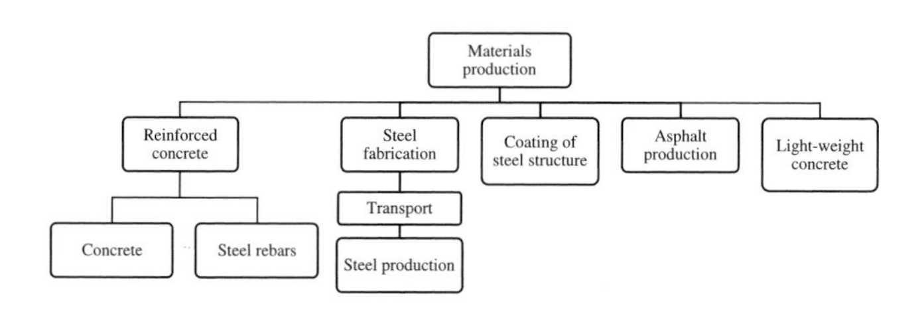

Figure 6.7 *Processes included in the stage of material production.*

Table 6.8 *Sources of data for materials and transportation.*

Material/process	*Source*
Concrete (several grades)	Portland Cement Association (US) (Marceau et al., 2007)
Structural steel	Worldsteel organization (World Steel Life Cycle Inventory, 2002)
Reinforcement steel	Worldsteel organization (World Steel Life Cycle Inventory, 2002)
Painting	Ecoinvent database (Frischknecht et al., 2004)
Asphalt	Ecoinvent database (Frischknecht et al., 2004)
Transportation of steel plate	Ecoinvent database (Frischknecht et al., 2004)

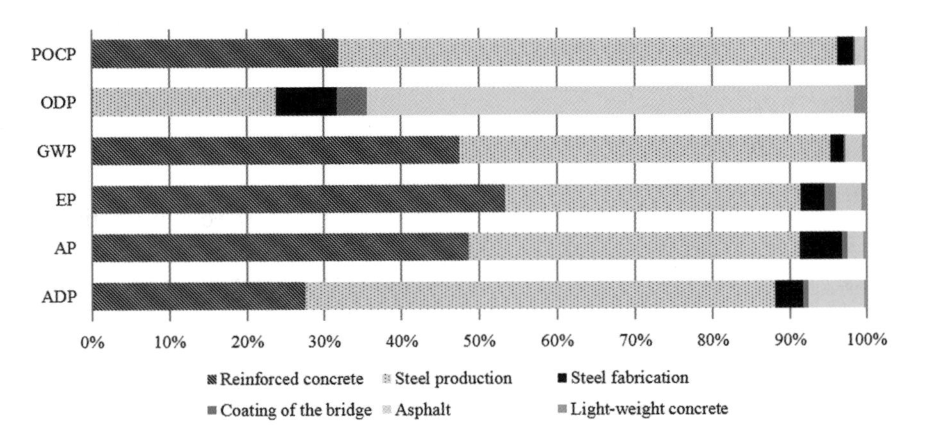

Figure 6.8 *Contribution analysis for the stage of material production.*

POCP – photochemical oxidation; ODP – ozone layer depletion; GWP – global warming; EP – eutrophication; AP – acidification; ADP – abiotic depletion (see text also).

6.3.4.1 Stage of material production

This stage takes into consideration the production of all the materials needed to build the bridge, according to Figure 6.7. Data were collected from several sources as indicated in Table 6.8.

According to information from the Portuguese Association for Steel and Composite Construction (CMM), most of the steel used in Portugal for construction (steel profiles and plates) is imported from Europe (Luxemburg, Spain, France, Italy and Germany). Furthermore, due to its proximity to Portugal, Spain is the first source of steel to Portugal. Hence, an average distance of 700 km was considered.

The results obtained in this stage are plotted in Figure 6.8, in terms of a contribution analysis in order to highlight the most important processes. The burdens due to the transportation of steel plate were included in the process of steel production. From Figure 6.8, it is concluded that the production of reinforced concrete and steel are the most significant processes for most impact categories in the stage of material production.

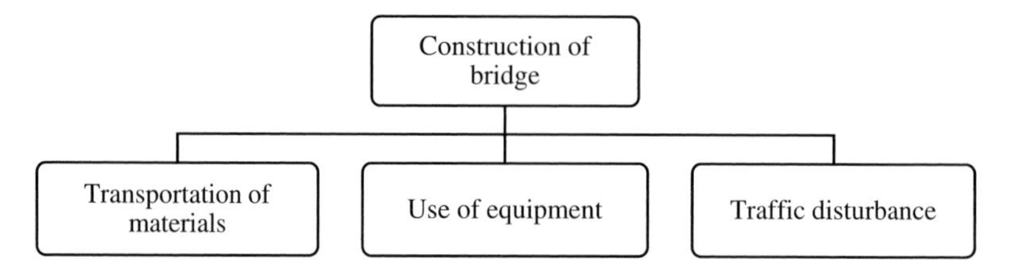

Figure 6.9 *Processes included in the stage of construction.*

Table 6.9 *Transportation of materials and equipment for the construction stage.*

Activity	*Distance (km)*
Transportation of soil and debris	10
Transportation of steel beams	50
Transportation of precast slabs	50
Transportation of fresh concrete	10
Transportation of steel reinforcement	50

6.3.4.2 Stage of construction

The construction stage takes into account all the processes needed for the construction of the bridge and affected by it. Hence, as represented in Figure 6.9, it includes the transportation of materials to the construction site (according to the distances indicated in Table 6.9) and the use of construction equipment. In addition, it takes into account also emissions due to the traffic congestion resulting from construction activity.

Transportation of materials and construction equipment

Construction materials have to be transported to the construction site. Some of the distances were calculated based on information from the constructor. However, most data was estimated due to lack of information. The travelling distances estimated for each case are indicated in Table 6.9 and the consumption of diesel is based on the travel distances displayed and the number of trips needed for each case.

The environmental data for fuel consumption of trucks and respective fuel production were taken from the ECOINVENT database (Frischknecht *et al.*, 2004).

Use of construction equipment

The construction of the bridge requires the use of heavy equipment such as crawler-excavators, loaders, bulldozers, water trucks, concrete mixers, cranes and so on, which are usually diesel powered.

The quantification of emissions due to the use of construction equipment is based in the plan of construction represented in Table 3, in the list of equipment provided by the constructor and the emission factors from the NONROAD model (EPA, 2003).

Traffic over and under the bridge

As previously referred, being a new bridge, only emissions from traffic beneath the bridge are taken into account. During the construction period, traffic is affected due to restrictions in the traffic speed and the narrowing of the carriageway. Traffic congestion due to work activity in the surrounding area of the bridge has two major types of impacts: (1) the impacts due to direct emissions from vehicles, and (2) the impacts due to the amount of fuel consumed. The impacts due to direct emissions from vehicles are quantified based on the QUEWZ-98 model. The impacts due to fuel production, including the upstream burdens, are quantified based on data from Ecoinvent. In both cases, the quantification of impacts takes into account only additional burdens that are the difference between the impacts of the vehicles passing through the work zone and the impacts of the vehicles passing through the same zone but without any work activity.

Results for the construction stage

The results obtained for each impact category are represented in Figure 6.10. In this stage, the most significant process is clearly "traffic congestion" in work zones. Its

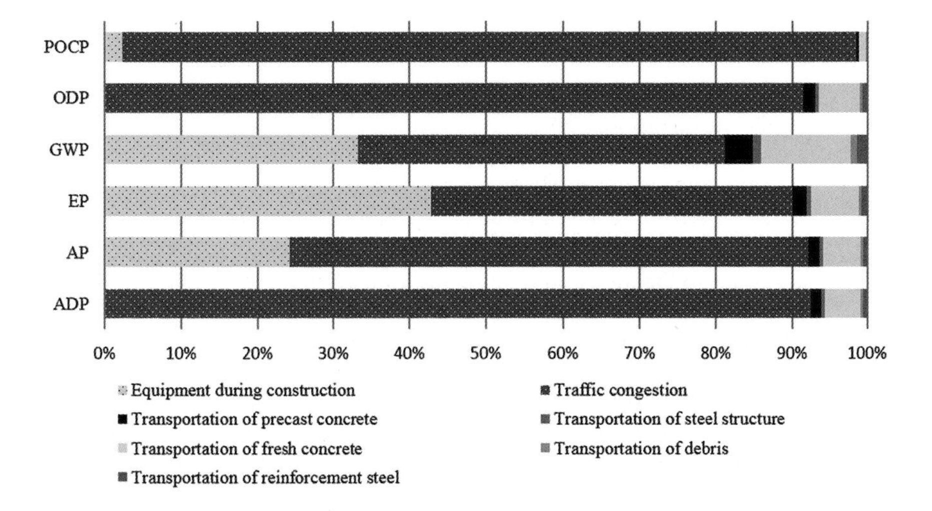

Figure 6.10 *Contribution analysis during the construction stage.*

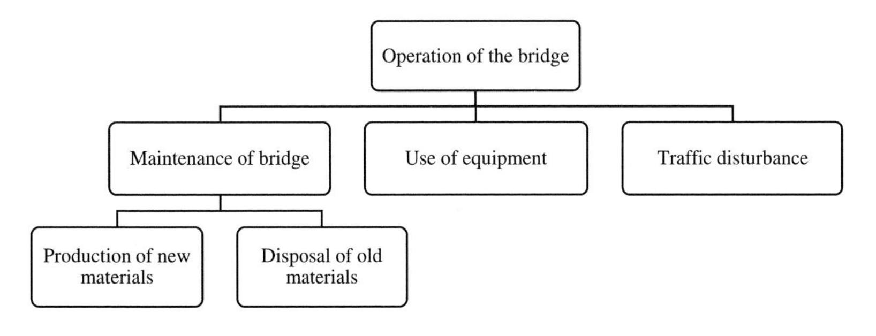

Figure 6.11 *Processes included in the stage of operation.*

contribution is always above 80%, except for the impact categories of acidification (AP), eutrophication (EP) and global warming (GWP), for which the "use of equipment" has also an important contribution.

6.3.4.3 Stage of operation

The operation stage takes into account all the processes needed for the maintenance of the bridge and affected by it, as represented in Figure 6.11. In some cases, during the maintenance of different bridge components, new materials are needed to replace old ones. In this case, the production of new materials and the disposal of old ones are also taken into account.

Transportation of materials and equipment

Each time the bridge undergoes an activity of maintenance or rehabilitation, according to Table 6.5, materials have to be transported to the bridge site and construction equipment is used. The travelling distances considered in this stage are the same as in the construction stage, unless otherwise indicated.

Use of equipment

The number of hours needed for equipment were calculated based on the time estimated for each activity and in the work capacity of each equipment (assumed that it was used at full capacity). Then, the NONROAD model (EPA, 2003) was used for the quantification of emissions.

Traffic over and under the bridge

During the operation stage, traffic is affected over and under the bridge. Whenever it is required to close one lane, two different scenarios may be considered: during the day period (from 0800 to 1700) and during the night period (2000 to 0600). Naturally,

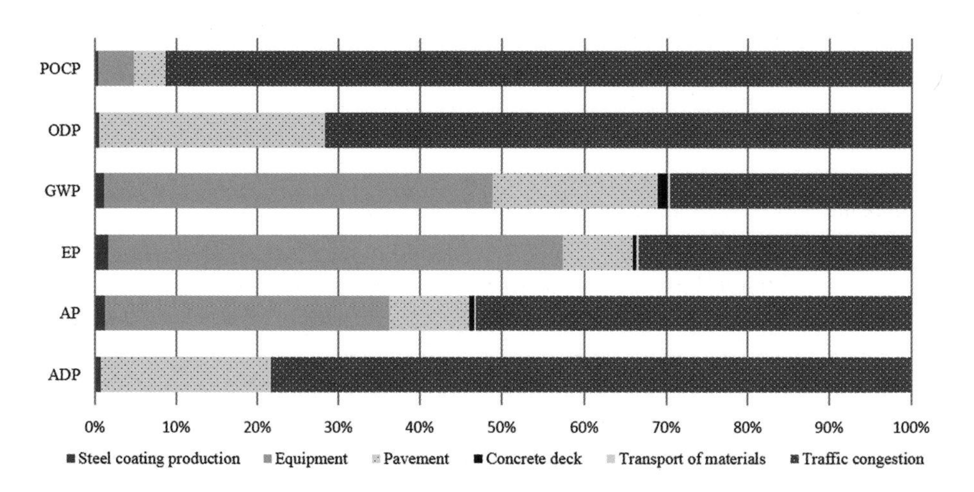

Figure 6.12 *Contribution analysis of processes during the operation stage.*

when the closure of the lane is made over the night period, traffic affection and respective burdens are minimized. Therefore, this was always the scenario considered, except for the following maintenance actions, which required the closure of one lane for the 24 hour period:

- Maintenance of concrete deck
- Maintenance of steel structure (replacement of coating system)
- Replacement of the asphalt layer

Results for the operation stage

The compiled results, for the entire stage of operation, are represented in Figure 6.12. Similar to the construction stage, the most significant process is "traffic congestion" in work zones. Its contribution is always above 80%, except for the impact categories of AP, EP and WP, for which the "use of equipment" has also an important contribution. It is also observed a significant contribution resulting from the maintenance/replacement of the asphalt layer of the bridge in the impact categories of ozone layer depletion (ODP), global warming and abiotic depletion (ADP).

6.3.4.4 Stage of decommission and final disposal

The stage of decommission takes into account the demolition of the bridge, the disposition of waste materials in their final destiny and related processes, as represented in

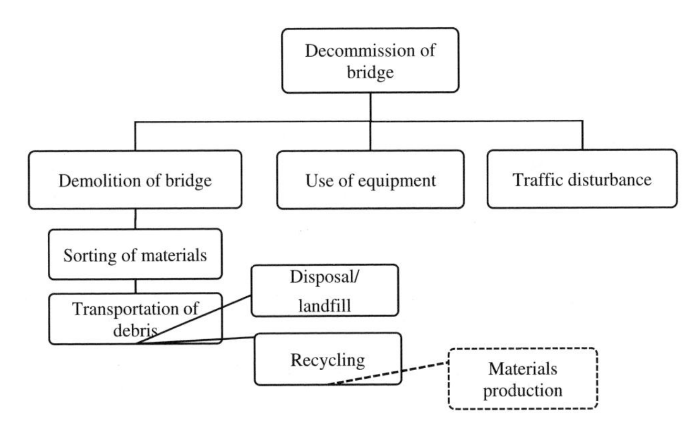

Figure 6.13 *Processes included in the stage of decommissioning.*

Table 6.10 *Transportation of materials for the end-of-life stage.*

Activity	*Distance (km)*
Deposition of soil	10 km
Landfill of inert materials	50 km
Recycling plant of steel reinforcement	50 km
Recycling plant of structural steel	50 km
Recycling plant of asphalt	20 km

Figure 6.13. The final destination of each material depends on its inherent characteristics, such as its potential for recycling or eventual reuse.

In this case study, it was considered that structural steel is recycled, with a recycling rate of 85%. The recycling of steel avoids the production of new steel via the blast furnace route, thus all environmental burdens associated with this route may be deducted from the analysis. Therefore, in this case, a closed material loop recycling approach is considered (World Steel Life Cycle Inventory, 2002). The same approach was considered for the recycling of steel reinforcement. In this case, a recycling rate of 70% was considered.

For the recycling of asphalt a cut-off rule was used due to the lack of data in relation to the respective recycling process. In this case, the only credit considered in the analysis is due to the reduction of waste sent to a landfill. All remaining materials were sent to a landfill of inert materials.

Transportation of materials and equipment

It is assumed that the bridge is demolished and the resulting materials are sorted in the demolition place. After sorting, materials are sent to their final destination according to the respective end-of-life scenario. The estimated travelling distances between the sorting place and the final destination of the materials are indicated in Table 6.10. Furthermore, it is assumed that the transportation of materials is done by truck.

Use of equipment

The quantification of emissions due to the use of demolition equipment is based in the plan of demolition represented in Table 6.7, in the estimated list of equipment and the emission factors from the NONROAD model (EPA, 2003).

Traffic over and under the bridge

During the demolition of the bridge, it is assumed that traffic over the bridge is diverted to an alternative road or is already being accommodated by an alternative

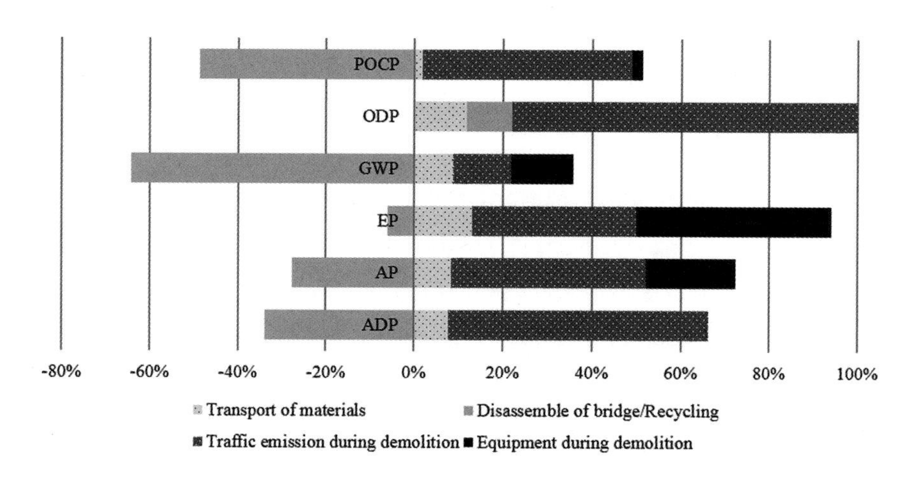

Figure 6.14 *Contribution analysis of processes during the end-of-life stage.*

bridge. In any case, no emissions and fuel consumption were considered for this traffic. Therefore, in the following sections fuel consumption and emissions are only accounted for traffic under the bridge, as it is assumed that the motorway is still in service by this time.

The demolition period is estimated to occupy 23 days. During this period, it is estimated there would be always three lanes of traffic in each direction (one lane closed), with reduced speed, save for four working days. These four working days correspond to the demolition of the central span of the structure. The central span is divided into two parts. Each part needs 2 days (8 hours per day) to be demolished. During this period, the corresponding carriageway has to be closed to the traffic underneath. During the closure of a carriageway, traffic has to be diverted to the other carriageway, resulting in three or, in the worst case, two lanes of traffic in each direction.

Results of the end-of-life stage

The results for the end-of-life stage are represented in Figure 6.14. The process of "traffic congestion" has a significant contribution to most impact categories, except GWP. The recycling of materials after the demolition plays also a significant role in

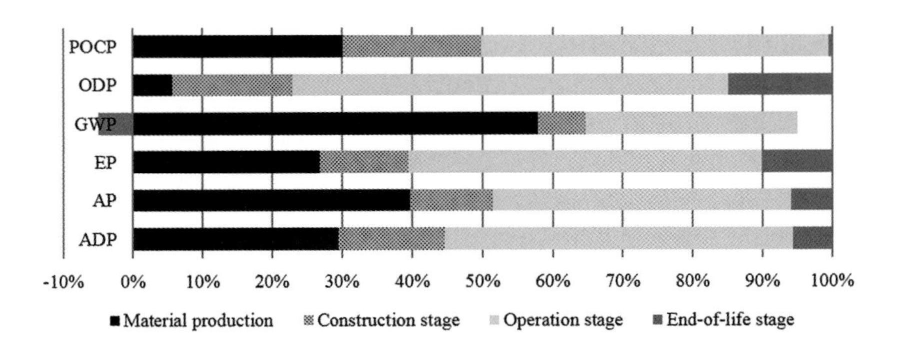

Figure 6.15 *Contribution of each stage to impact category.*

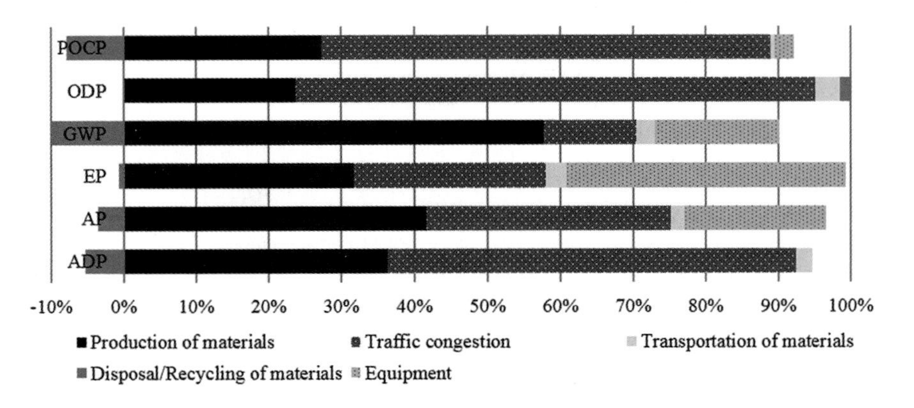

Figure 6.16 *Contribution of each main process to impact category.*

this stage, particularly for the impact category of global warming. The negative values in Figure 6.14 represent the credits given to the recycling processes.

6.3.4.5 Synthesis of life cycle environmental analysis

The contribution of each life cycle stage to each environmental category is illustrated in Figure 6.15. The stage of material production has a significant contribution to most impact categories, in particular for GWP with a percentage above 50%. On the other side, this stage has a minimum contribution to the impact category of ODP, with a percentage less than 10%. The stage of operation stage has a major contribution (close to 50%) to most impact categories except for GWP.

The stage of construction has a contribution less than 20% for all impact categories; while, the end-of-life stage has a global contribution of less than 10%, except for ODP.

The results presented in Figure 6.15 were rearranged according to the main process involved in the life-cycle analysis and the results are represented in Figure 6.16. Five main processes were identified: (1) production of materials, (2) transportation of materials and equipment, (3) use of equipment, (4) traffic congestion and (5) disposal/recycling of materials.

According to Figure 6.16, the processes of production of materials and traffic congestion are dominant for most impact categories. The production of materials has a contribution above 50% for the impact category of GWP; while traffic congestion has a major contribution (above 50%) for the impact categories of photochemical oxidation (POCP), ODP and ADP. The process of use of equipment has a significant contribution to the impact categories of EP, GWP and AP. Finally, the process of disposal and recycling of materials has a contribution less than 10% for all impact categories and transport has, in general, a negligible contribution.

From the previous results, it is concluded that the production of materials and traffic congestion in work zones are the most critical issues in the life cycle of the bridge.

6.3.5 Life-cycle cost analysis

The scope of life cycle cost analysis (LCCA) is shown in Figure 6.2. Therefore, direct costs are quantified taking into account the construction, maintenance and demolition of the bridge. In this case, costs due to construction were calculated based on the

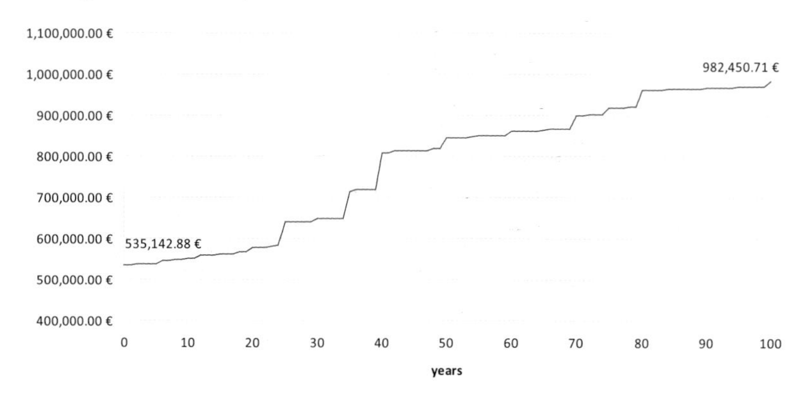

Figure 6.17 *Life cycle cost of the bridge (direct costs).*

bill of materials provided by the contractor and respective unit costs, leading to a value of 535,143.00 €.

Costs due to the maintenance of the bridge over its life cycle, were estimated based on the maintenance scenario and unit costs indicated in Table 6.5. These costs encompass the cost of labour work, the cost of equipment, the cost of road warning signage and the cost of transportation. In addition, end-of-life costs include the cost for deposition of materials and/or revenue due to recycling of materials. In this case, a total cost of 81,184.00 € was obtained (Gervásio, 2010), based on information from the contractor, on an estimated cost for scrap revenue (100 €/ton) and on the estimated fees for disposal of different materials.

Hence, the compilation of life cycle costs leads to the total present value of 982,450.00 €, which represents a cost of about 1,048.00 €/m². The present value was obtained from expression 6.2 considering a discount rate of 2.0%. The accumulated present value of the bridge is illustrated in Figure 6.17.

6.3.6 Life cycle user costs

User costs are computed according to the methodology described previously. For the quantification of user costs, the scenarios described for life cycle environmental analysis (LCEA) were taken into account. It is noted once again that during the construction

Table 6.11 *DTC per hour and class of vehicle (DTC_i).*

Class of vehicle	*Class 1*	*Class 2*	*Class 3*	*Class 4*
DTC_i (€/h)	7.75	6.20	62.90	9.30
VOC_i (€/km)	0.17	0.12	0.83	0.67

Table 6.12 *Cost of assistance per type of accident (ca_j) and cost per victim (cv_k).*

Type of accident	*j*	*Cost of accident (ca_j)*	*Type of victim*	*k*	*Cost of victim (cv_k)*
With light-injuries	1	73.12 €	Light-injury	1	41,600 €
With severe-injuries	2	255.31 €	Severe-injury	2	93,600 €
With fatalities	3	342.67 €	Fatality	3	520,000 €

Table 6.13 *Rates under normal traffic conditions and in work zones (x 10^{-8}).*

	Rates under normal traffic conditions (RAn_j and RVn_k)			*Rates in work zones traffic conditions (RAaj and RVak)*		
Type j = k	1	2	3	1	2	3
Rate of accident type j	11.54	1.05	0.50	12.12	1.10	0.53
Rate of victims type k	17.36	1.46	0.58	18.23	1.53	0.61

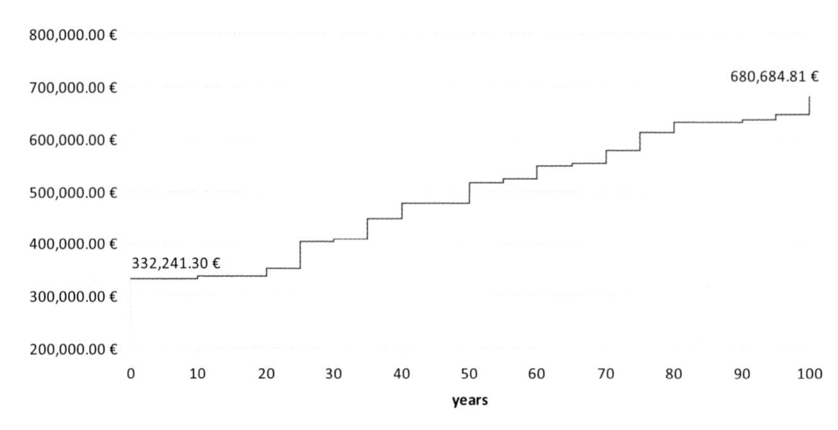

Figure 6.18 *Life cycle cost of the bridge (users' costs).*

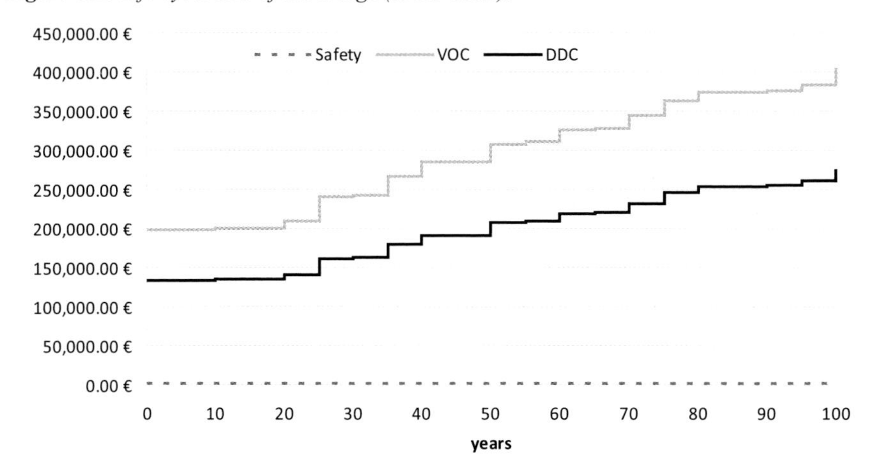

Figure 6.19 *User's costs—partial values.*

of the bridge and during its demolition (according to the scenario considered), the highway was in service although with traffic constraints, as mentioned before.

Additional data are needed to quantify user costs. The quantification of DDC and VOC takes into account four classes of vehicles: (1) light-weight vehicles (class 1); (2) vehicles with two axes and height equal or above 1.10 m (class 2); (3) vehicles with three axes (class 3); and (4) vehicles with more axes (class 4).

Hence, the cost per hour of a driver's time of a class-i vehicle (DTC_i) and the operation costs of a class-i vehicle (VOC_i), per km are indicated in Table 6.11, for the different classes of vehicles (Gervásio and Simões da Silva, 2013b).

In relation to accident costs, three types of main accidents are considered (1) accidents with light-injuries, (2) accidents with severe-injuries and (3) accidents with fatalities. The costs per type of accidents and the respective cost of victims are indicated in Table 6.12 (Gervásio and Simões da Silva, 2013b).

The rate of accidents and the rate of victims, under normal traffic conditions and in work zones, are indicated in Table 6.13, per type of accident (j) (Gervásio and Simões da Silva, 2013b).

Hence, the compilation of user costs over the service life of the bridge lead to the total present value of 680,685.00 €, which represents a cost of 727.00 €/m² and is about 69% of the total direct cost. The same discount rate was considered (2.0%) for the calculation of the present value. The accumulated present value of the bridge is illustrated in Figure 6.18.

User costs comprehend three components: DDC, VOC and the cost of safety. The accumulated present value of each component is illustrated in Figure 6.19.

Contrary to other studies (Wilde *et al.*, 1999; Daniels *et al.*, 1999), VOC is the most significant component (59.3%) followed by DDC (40.4%). The main reasons for this may be the higher cost of the fuel (which is reflected in VOC_i) and lower incomes in Portugal (which is reflected in DDC_i), in comparison with other countries.

Safety costs are negligible. This was already expected from Table 6.13, since due to safety measures currently in highways, accident rates in work zones are close to accident rates in nonwork zones (Gervásio, 2010).

6.3.7 Synopsis of the case study

The results of this case study highlighted the importance of performing LCA. Ignoring the impacts and costs occurring over the life cycle of the bridge leads to a significant underestimation of the problem. Furthermore, it is once more observed that for the computation of environmental burdens and user costs, the best scenarios were always considered. This may be a too optimistic perspective leading to an underestimation of the results.

Given the importance of the operation stage, namely the environmental burdens and user costs due to traffic congestion during construction activity, an alternative case study is provided in the next subsection aiming for an improvement in the life cycle performance of the bridge.

6.3.8 Alternative case study

An alternative design solution is herein presented, which consist in the replacement of the original carbon steel by weathering steel, avoiding the use of a coating system.

Weathering steel is a high strength low alloy steel that may be left unpainted as it forms an adherent protective rust "patina" to prevent further corrosion. In a stable environment, the life of a weathering steel bridge can be more than 100 years (Zaki *et al.*, 2005).

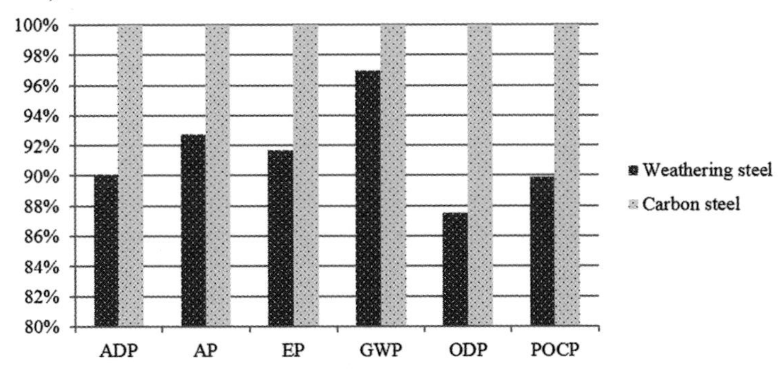

Figure 6.20 *Comparison of life cycle environmental analysis.*

Therefore, in this case study it is assumed that during the life cycle of the bridge there is no need for the maintenance of the steel structure. This is a realistic assumption provided that the bridge is periodically inspected and cleaned. The rusted material has no contribution to the structural capacity of the steel element. To enable the loss of section in a weathering steel section, the design of such section requires an additional thickness, which depends on the severity of the environment. Assuming a medium environment, an additional thickness of 1.5 mm in both faces was considered for all steel sections (Zaki *et al.*, 2005), leading to an additional weight of the steel structure of about 10%.

Currently, there is no inventory data of weathering steel products in commercial databases. Therefore, the same environmental burdens due to the production of 1 kg of a steel section and 1 kg of a steel plate are considered.

Likewise, all the remaining data and scenarios in this case study coincide with the previous analysis. Hence, the results of the LCEA are compared with the results of the previous analysis in Figure 6.20. Although there was an additional burden in the initial stage due to the production of more steel, the elimination of the coating system and, in particular, the avoidance of the maintenance of the steel structure led to reductions between 3 and 12%. The higher reduction was for the environmental category of ODP

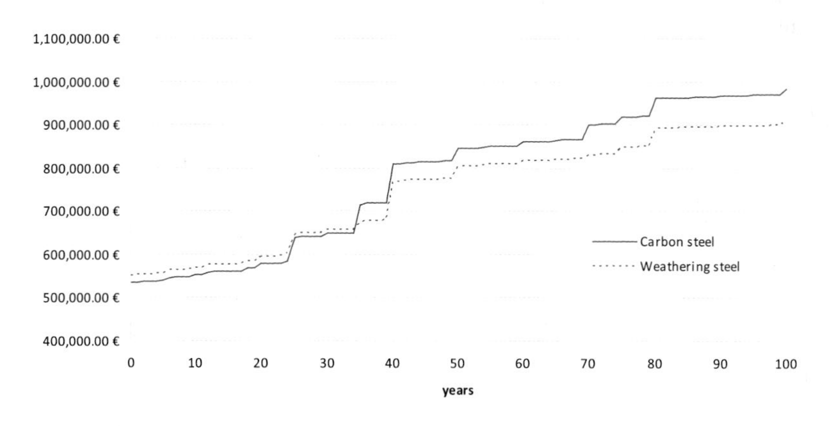

Figure 6.21 *Life cycle cost of the bridge (direct costs).*

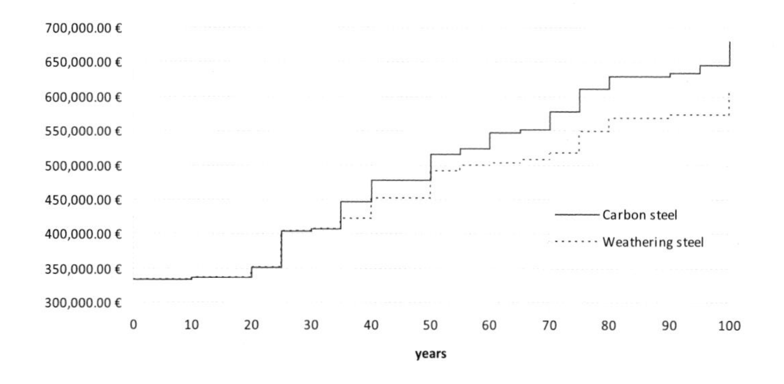

Figure 6.22 *Life cycle user costs of the bridge.*

due to the contribution of the coating in this category, and the lower for GWP, as the production of steel is a major contributor for CO_2 emissions.

The material cost of weathering steel is greater than of carbon steels. According to information received from a major steel producer, an additional 80 € per tonne of steel is usually a reference value. Taking this value into account, the results of LCCA, taking into account direct costs, are represented in Figure 6.21.

In Figure 6.21, the results of the present analysis are compared with the results of the previous analysis. Although, the initial cost of construction is higher, taking into account a life cycle perspective, the savings due to the elimination of the coating system and respective maintenance/replacement clearly outweighed the additional material cost. Similarly, avoiding the maintenance and replacement of the coating systems and related traffic affectation, it is enough to reduce users' costs over the bridge's life cycle, as observed from Figure 6.22.

Once again this simple example illustrates the importance of LCA. In this particular case, focusing only on the initial stage may lead to erroneous conclusions, in that a solution, which can be more expensive initially, turns out to be more advantageous in the long term.

In addition, this example shows that by minimizing the maintenance needs of the bridge over its life cycle leads to a substantial improvement of the lifetime performance of the bridge, particularly when environmental impacts and user costs are taken into account.

6.4 Conclusions

This chapter describes the application an integral life cycle approach to a composite bridge, considering the environmental, economic and social criteria.

The aim of the integrated life cycle approach is to assess the life cycle performance of the bridge, enabling the identification of the stage(s) and process (es) with major impacts and the causes of those impacts. This has to be accomplished by a comprehensive data collection and impact evaluation in terms of the three main aspects.

From the environmental analysis, it was concluded that the stages of material production and operation are the ones with major importance. In terms of processes involved in the analysis, the production of materials and traffic congestion are the most important ones, with major contribution to all impact categories. From this analysis, it may be concluded that the use of construction materials with lesser environmental impacts and a well-planned management of construction work throughout the life cycle of the bridge are important factors to consider in the pursuit of an improved life cycle environmental performance.

The results of the LCCA and LCSA allow to conclude about the importance of minimizing maintenance and rehabilitation activities over the service life of the structure. Furthermore, in the case study, the total user cost was close to 70% of the total direct cost of the bridge, assuming optimistic scenarios. This shows that apart from direct costs borne by highway administrations, the costs supported by the users and environmental impacts are very significant over the lifetime of the bridge and should not be neglected. Unfortunately, this is not the current practice in most cases.

References

Daniels, P., D. Ellis and W. Stockton (1999) *Techniques for manual estimation road user costs associated with construction projects*. Texas Transportation Institute, Texas, USA, Dec.

EN 1990 (2002) *Eurocode—Basis of structural design*. European Committee for Standardization, Brussels, Belgium.

EPA (2003) *NONROAD model, 2002a*. U.S. Environmental Protection Agency Washington, D.C.

Frischknecht, R., N. Jungbluth, H.-J. Althaus, G. Doka., R. Dones, T. Heck, S. Hellweg, R. Hischier, T. Nemecek, G. Rebitzer and M. Spielmann (2004) *The ecoinvent database: overview and methodology, CD-ROM*, Final Report Ecoinvent 2000 No. 1. Swiss Centre for Life Cycle Inventories, www.ecoinvent.ch, Dübendorf, CH.

Fuller, S. and S. Petersen (1996) *Life-cycle costing manual for the Federal energy management program*, NIST Handbook 135, 1995 Edition. Building and Fire Research Laboratory, Office of applied economics, Gaithersburg, MD 20899.

Gervásio, H. (2010) *Sustainable design and integral life-cycle analysis of bridges*, PhD Thesis. University of Coimbra, Coimbra, Portugal.

Gervásio, H. and L. Simões da Silva (2013a) A design approach for sustainable bridges—Part 1: Methodology. In: *Proceedings of ICE—Engineering Sustainability* **166** (4), 191–200, DOI information: 10.1680/ensu.12.00002.

Gervásio, H. and L. Simões da Silva (2013b) Life-cycle social analysis of bridges. *Journal of Structure and Infrastructure Engineering* **9**(10), 1019–1039, http://dx.doi.org/10.1080/15732479.2011.654124.

Guinée, J., M. Gorrée, R. Heijungs, G. Huppes, R. Kleijn, A. de Koning, L. van Oers, A. Sleeswijk, S. Suh and H. Udo de Haes (2001) *Life cycle assessment: an operational guide to the ISO standards—part 2b—operational annex*. Centre of Environmental Science (CML), Leiden University, Leiden, Netherlands.

ISO 14040 (2006) *Environmental management—life cycle assessment—principles and framework*. International Organization for Standardization, Geneva, Switzerland.

ISO 14044 (2006) *Environmental management—life cycle assessment—requirements and guidelines*. International Organization for Standardization, Geneva, Switzerland.

ISO 15686-5 (2008) *Buildings and constructed assets—Service-life planning —part 5: life-cycle costing*. International Organisation of Standardisation, Geneva, Switzerland.

Marceau, M., M. Nisbet and M. VanGeem (2007) *Life cycle inventory of Portland cement concrete*, PCA R&D Serial No. 3011. Portland Cement Association, Skokie, Illinois.

SBRI (2013) *Sustainable steel-composite bridges in built environment*, Draft Final report, RFSR-CT-2009-00020. Research Programme of the Research Fund for Coal and Steel (available from http://bookshop.europa.eu/en/sustainable-steel-composite-bridges-in-built-environment-sbri--pbKINA26322/) European Commission, Brussels.

Seshadri, P. and R. Harrison (1993) *Workzone mobile source emission prediction*, SWUTC/92/60021-3. Centre for transportation research, University of Texas at Austin, Austin, Texas.

SimaPro 7 (2008) *Software and database manual*. PRé Consultants, Amersfoort, Netherlands.

Wilde, W., S. Waalkes and R. Harrison (1999) *Life cycle cost analysis of Portland cement concrete pavements*, Report no. FHWA/TX-00/0-1739-1. Centre for Transportation Research, Austin, Texas, USA.

World Steel Life Cycle Inventory (2002) *Methodology report 1999/2000*, International Iron and Steel Institute. Committee on Environmental Affairs, Brussels, Belgium.

Zaki, R., C. Clifton and C. Hyland (2005) *New Zealand weathering steel guide for bridges*, HERA Report R4-97:2005. New Zealand Heavy Engineering Research Association, Manukau City, New Zealand.

Chapter 7

Life cycle cost analysis for corrosion protective coatings—offshore wind turbines

Astrid Bjorgum, Thomas Welte and Matthias Hofmann

An analysis of life cycle costs (LCC) of corrosion protective coatings for offshore wind turbines is presented in this chapter. The aim of the analysis is to compare different coating systems and their influence on the costs. Since the main focus is on comparing different coating systems and to find the most cost effective coating solution, a simple approach for life cycle cost analysis (LCCA) is used where the focus is on cost elements that are influenced by the choice of coating system. Thus, other cost elements (i.e. cost elements that are not influenced by the choice of coating system, or cost elements that are considered to be negligible) are not taken into account in the LCCA.

In the first section, a short introduction to offshore wind turbine structures is given, and requirements and challenges regarding corrosion protection are discussed. An overview of experience in corrosion protection from the offshore oil and gas industry and Norwegian road bridges is given in Section 7.2 with guidelines and recommendations for offshore wind power installations. Coating systems for coating repair are discussed in Section 7.3. Section 7.4 presents typical costs for coatings and coating application. Areas that need to be protected and their size are discussed in Section 7.5. The LCCA and the results and conclusions from this analysis are presented in sections 7.6 and 7.7.

7.1 Introduction

An offshore wind turbine consists basically of the following parts: foundation, substructure, transition piece with platform, tower and the rotor/nacelle assembly. An illustration of an offshore wind turbine for a typical design with monopile foundation is shown in Figure 7.1. The combination of foundation, substructure, transition piece and tower are defined as the support structure (Quarton, 2005) for the nacelle assembly. Design of offshore constructions with long lifetimes needs to consider the environmental conditions at sites and thus the selection of the corrosion protection systems to be used. Corrosion protection consists of protective coating systems and in submerged areas cathodic protection. The external surfaces of steel structures are generally divided in three zones differing with respect to corrosivity and corrosion protection methods. These are the atmospheric zone (AZ), splash zone (SPZ) and submerged zone (SMZ).

In the SMZ, cathodic protection using sacrificial anodes will normally give satisfying corrosion protection. Sacrificial anodes in combination with a paint system are proved to give the desired corrosion protection for the lifetime of oil- and gas

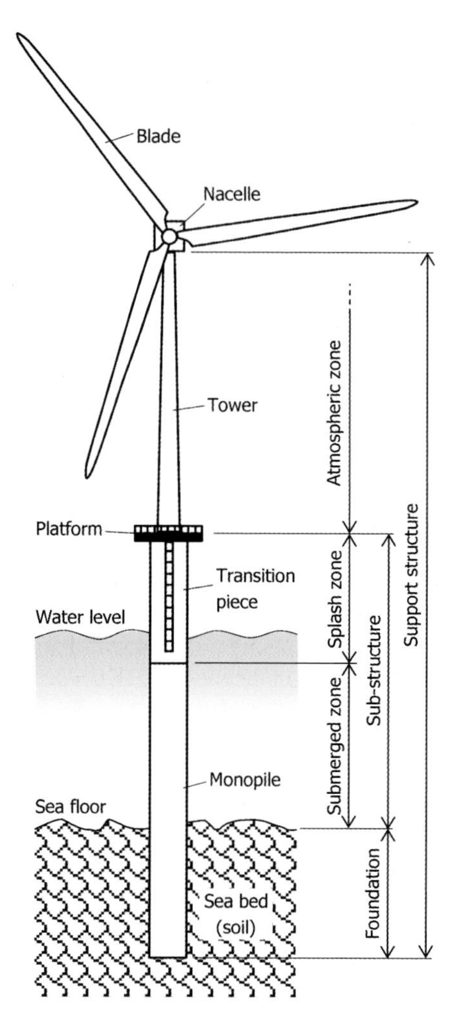

Figure 7.1 *Main parts of an offshore wind turbine illustrated by a turbine with monopile substructure and foundation.*

installations, i.e., 30–50 years. In the AZ and the SPZ, coating systems need to be applied to protect steel against corrosion. SPZ is that part of the offshore installation which is intermittently exposed to air and immersed in the seawater. Due to the constant presence of seawater and oxygen, the SPZ is demanding with respect to corrosion. The corrosivity is significantly higher than in the AZ, and the steel surfaces in this zone therefore need extra corrosion protection. Cathodic protection will not work since the steel structure is not continuously immersed in the seawater. Thus, the steel structure should be protected by a coating system proven suitable for the harsh environmental conditions in this zone. In addition a corrosion allowance (i.e. an extra thickness) is normally added to the steel in the SPZ as shown, e.g. in NORSOK M-001 (2004).

Internal steel surfaces above and below the water surface will also need corrosion protection. Corrosion protection inside steel support structures may be achieved by coating application and/or dehumidification. Internal climate control devices may also be used. Available information (van der Mijle Meijer, 2009; Herion *et al.*, 2009) shows that corrosion occurs inside sealed wind turbine foundations although limited oxygen contents are present. Microorganisms from the seabed may result in microbiological induced corrosion (MIC). Hence, in oxygen free areas like the seabed and internal water filled voids, an increased corrosion rate is caused by the microorganisms acting as catalysts for the corrosion process.

Existing experience from offshore oil and gas installations has shown that coating maintenance almost always is demanded to obtain a long lifetime for equipment and structure in the AZ. Offshore coating repair is very expensive due to climate and weather conditions, and the reduced access and operation windows compared to coating maintenance onshore. In addition, the environmental conditions will affect the quality of the coating repair work, resulting in shorter lifetime than for the original coating applied in yards. We have seen that introduction of new coating systems carries a risk of premature coating failure (Knudsen *et al.*, 2004) and should only be done after careful documentation of the new system. Repair and maintenance costs need to be taken into consideration when selecting coating systems for new constructions. Despite higher investments costs, coating systems with longer lifetime that requires less maintenance may be beneficial in the long run (Knudsen *et al.*, 2012).

7.2 Protective coating systems

7.2.1 Experience from oil and gas installations

When building new offshore oil and gas installations, selection of corrosion protective coating systems is performed in accordance to international (ISO 12944-5, 2007; ISO 20340, 2009; NORSOK M-501, 2012) and company specific standards. ISO 12944-5 (2007) describes paint systems used for corrosion protection and provides guidance on coating systems to be used under different environmental conditions. In the Norwegian sector, protective coating systems for offshore oil and gas installations are usually selected in accordance with NORSOK M-501 (2012). This standard specifies functional requirements for coatings, surface preparation prior to coating application, generic type of coatings, film thicknesses and number of coats, as well as application procedure and inspection for the protective coatings, to be applied during construction and installation on site. NORSOK M-501 refers to ISO 20340 (2009) for prequalification of the coating systems. Table 7.1 shows typical coating systems used in the AZ, SPZ and SMZ of an offshore oil and gas installation, as well as lifetime expectancy based on experience from the past 20 years. Long lifetimes are experienced for corrosion protection systems applied in the SPZ and SMZ. For the coating systems used in the AZ, however, 8–10 years maintenance intervals are experienced. For the Statfjord field, one of the oldest oil and gas fields in the North Sea with a coated area of about 1.2 million square meters, the coating maintenance interval exceeded about 12 years for 95% of the coated area (Axelsen *et al.*, 2010). It is emphasized that Statfjord is one of the fields allowing the highest coating degradation before refurbishment.

Table 7.1 *NORSOK M-501 (2012) recommended coating systems for offshore installations.*

Exposure conditions	*Typical coating system*	*Lifetime expectancy*
Atmospheric zone	Zinc epoxy: 60 µm Epoxy barrier coat: 150 µm UV resistant topcoat: 70 µm	Time to first major maintenance is normally about 10 years. Typical maintenance intervals about 12 years (Axelsen *et al.*, 2010)
Splash zone	2-coats polyester Mean dry film thickness: >1000 µm	Lifetime of 20 years or more is usually achieved
Submerged zone	2-coats epoxy Mean dry film thickness : 350 µm	According to design life, degradation of the coating is compensated by sacrificial anodes

7.2.2 Experience from road bridges

Since 1965, the Norwegian Public Roads Administration (NPRA) has used so called duplex coating systems, consisting of a thermally sprayed metal layer with a paint system on top, to protect steel bridges (Knudsen *et al.*, 2012). The duplex systems they have used consist of 100 µm thermally sprayed zinc (TSZ), a sealer coating, and 2–4 layer paint system, 200–300 µm in thickness. NPRA has experienced long lifetimes and little or no need for maintenance of duplex coatings applied on coastal bridges. One example of successful long life performance is the Rombak Bridge crossing the Rombak fjord in northern Norway. No corrosion was seen after 39 years in service (Klinge, 1999, 2009). Thermally sprayed metallic coating systems are also preferred as corrosion protection of bridges in other European countries, and in the United States. In the United Kingdom more than 90% of all new steel bridges are protected by duplex coating systems (CORRPRO, 2001).

7.2.3 Recommendations for offshore wind turbines

Corrosion and coating schemes related to the design basis for offshore wind turbine support structures is discussed in Petersen *et al.* (2005). DNV-OS-J101 (2011) has also issued a standard for offshore wind turbine structures. To ensure a lifetime of the coating system that corresponds to the design life of the wind turbine (under the assumption that the coating is only subject to minimum maintenance), DNV recommends application of coating systems with documented performance, either by prequalification in accordance with a standard or by operational experience. Furthermore, they recommend that the application process, including surface preparation (cleaning, grit blasting) and application conditions (relative humidity, temperature, coating thicknesses etc.), should be followed up and ensured to follow the selected rules or guidelines, e.g. NORSOK M-501. The NORSOK recommendations for corrosion protection methods in different zones (Table 7.1) are also included in the DNV-OS-J101 (2014) standard, see Table 7.2.

In Germany, design, manufacturing and installation of the corrosion protection systems for offshore wind turbines are subject to approval by DNV GL (2007, 2012).

Table 7.2 *DNV's recommendations for corrosion protection in the different zones of offshore wind turbines.*

Exposure conditions	*Typical coating system*	*Dry film thickness (DFT)*
Atmospheric zone	Coating system according to ISO 12944-5, category C5-M (very high corrosivity), typically - Zinc rich epoxy primer - Intermediate epoxy - Epoxy or polyurethane topcoat, polyurethane if a colour or gloss retention is required	Minimum 320 μm in total
Splash zone	Coating systems shall meet the requirements of NORSOK M-501 and ISO 20340 and be qualified for compatibility with cathodic protection systems. Consideration of loads of impacts from service vessels and floating ice should be taken. - Glass flake reinforced epoxy or polyester or - Thermally sprayed aluminium (TSA) with a silicon sealer	Minimum 1.5 mm Minimum 200 μm
Submerged zone	Coating below splash zone is optional. Qualification for compatibility with cathodic protection systems is necessary - Multilayer two component epoxy and cathodic protection - Alternatively cathodic protection only	Minimum 450 μm No coating

Data sources: Petersen *et al.* (2005) and DNV-OS-J101 (2014).

Before coating application, the steel surfaces should be prepared in accordance to ISO 12944-4 (1998) or an equivalent standard. Paint coatings or metallic coatings to be used should be selected according to ISO 12944-5 or an equivalent standard (DNV GL, 2013). Coatings proven to be suitable by practical use or well-founded testing can be used if accepted by DNV GL.

7.2.4 Summary

Based on the DNV-OS-J101 (2014) standard, the general demands for corrosion protection in different zones were summarized by Momber (2011), as shown in Table 7.3. In addition to protective coating systems and cathodic protection (in submerged areas), corrosion allowances demanded for structures in the SPZ and the scour zone at the seabed were included (see Table 7.3). ISO 12944, NORSOK M-501 and ISO 20340 are generally recommended standards for design, application and testing of protective coating systems for offshore wind turbines. Selection of coating systems for steel support structures with reference to the three standards are summarized by Mühlberg (2010) and shown in Tables 7.4 and 7.5. Application of a 2- or 3-coat paint system 240 μm in thickness is recommended for internal surfaces (*ibid.*).

Duplex coating systems, i.e. metallization combined with 3-coat paint systems are recommended (Momber, 2011) when high durability (lifetime above 15 years) is desired. The metallization is typically TSZ alloy consisting of 85% zinc and 15% aluminium (TSZA). The outer surface of the towers produced for offshore wind projects in Europe is often protected by a duplex system including

Table 7.3 *General demands on corrosion protection of offshore wind energy installations in the North Sea.*

Location	*Accessibility*	*Corrosion protection*	*Corrosion allowance*
Atmospheric	Yes	Coating system (C5-M)[a]	No
SPZ – splash zone	No	Coating system[b,c]	6 mm for stationary parts (20 years) 2mm for demountable parts
SMZ – submerged zone	No	Cathodic corrosion protection; Optional: coating system[b]	No
Scour zone	No	Cathodic corrosion protection; Optional: coating system[b]	3 mm
Below sea ground	No	None	None
Closed sections (with seawater)	No	Cathodic corrosion protection; Coating system in free areas	According to concrete conditions

Data source: Momber (2011).
[a]As per ISO 12944-1.
[b]Confirmation of CP-compatibility.
[c]Resistant against mechanical load/impact.

Table 7.4 *Coating systems for the atmospheric zone.*

Norm/standard	*Prime coat*	*Number of layers*	*Total dry film thickness in μm*
DIN EN ISO 12944	EP, PUR	3–5	320
C5-Marine, high	EP, PUR	2	500
	EP, PUR (zinc rich)	4–5	320
ISO 20340,	EP (zinc rich)	min. 3	>280
C5-Marine, high	EP	min. 3	>350
NORSOK M 501	EP (zinc rich)	min. 3	>280
	EP	min. 2	>1000

Data source: Mühlberg (2010).
Note that EP: Epoxy resin; PUR: Polyurethane resin; Zinc rich: Zinc rich primer, C5-Marine, high: Marine atmosphere, high corrosivity.

TSZA (60–80 μm in thickness) and a 3-coat paint system consisting of two epoxy coats (100–120 μm), and a polyurethane top coat (50–80 μm) (Mühlberg, 2010). Fast degradation and corrosion is experienced for thermally sprayed aluminium (TSA) covered with a thick paint system and should therefore never be used (Knudsen *et al.*, 2004, 2012). TSA with a thin sealer, however, is proven to give highly durable corrosion protection and is also a recommended protection in the SPZ of offshore structures,

Table 7.5 *Coating systems for the submerged and the splash zone.*

Norm/standard	*Prime coat*	*Number of layers*	*Total dry film thickness in μm*
DIN EN ISO 12944, Im2, high, (splash zone not described)	EP, (zinc rich)	3–5	540
	EP, PUR	1–3	600
	EP	1	800
ISO 20340, Im2, high, splash zone	EP, PUR (zinc rich)	min. 3	>450
	EP, PUR	min. 3	>450
	EP	min. 2	>600
NORSOK M-501*	EP	min. 2	min. 350

Data source: Mühlberg (2010).
Note that coating abbreviations are as in Table 7.4; Im2: immersed in seawater, high corrosivity.
*NORSOK M-501 demands that a coating system shall always be combined with cathodic protection.

as shown in Table 7.2. Metallization is usually not used inside the support structures. In some offshore wind turbine projects, metallization has been recommended for the lower part inside the towers (Mühlberg, 2010) and is also used (Momber, 2011) as internal corrosion protection in the SPZ of wind structures. According to Mühlberg (2010), offshore wind turbine structures may be less often metallized in the future due to the demands for less time spent on painting and cheaper surface treatments combined with good experience with high quality paint systems. However, duplex coating systems are still used as corrosion protection for wind turbine towers.

7.3 Coating systems for coating repair in atmospheric zone offshore

Temperature, relative humidity, wind speed and wave heights are determining factors for whether coating maintenance or repair work can be performed on offshore installations. According to NORSOK M-501 (2012), no final blast cleaning or coating application shall be done if the relative humidity is above 85% or when the steel temperature is less than 3°C above the dew point. Coatings shall only be applied or cured at ambient and steel temperatures above 0°C. NORSOK M-501, however, is only valid for the new construction phase, and not for coating maintenance. Prior to recoating, delaminated coating, dirt and salt need to be removed by some sort of surface preparation. This may be water jetting, grit blasting, a combination of water jetting and grit blasting or any kind of mechanical treatment. The steel surface needs to be dry before painting.

The existing repair paint systems should be applied in temperature and humidity windows defined by the coating suppliers. For coating systems consisting of two or more coats, the drying time before applying the second coat is important. The recoating interval depends on the temperature and the specific coating products in use. Both minimum and maximum recoating intervals needs to be observed indicating that one needs to ensure that the second coat can be applied within the given recoating time. For some coating systems, winter grade variants are available that can be applied at low temperature. In the LCCA that is presented in Section 7.6, we assume that these coatings can be applied for maintenance purposes.

7.4 Coating costs

The costs for coating application on offshore structures has been discussed by Mühlberg (2010) and a summary is shown in Table 7.6. Costs for coating application including all job-related costs, e.g. grit blasting and scaffolding, are included. The table shows that coating maintenance offshore may be more than 100 times more costly compared to coating application onshore, while maintenance of onshore wind turbines is about 5–10 times more expensive than coating application on new constructions in the paint shop.

Information received from Dong Energy (2012) also indicates that coating maintenance is probably 10 to 100 higher per square meter, depending on the exact conditions. Hence offshore maintenance is very expensive.

Safinah Ltd (Kattan, 2011) provides data for coating application costs in European shipyards. The type of coating systems corresponding to the costs were not reported, but paint systems used on ships are comparable to the ones used on offshore installations. The coating costs reported for new constructions were 10 €/m^2 on block and 125 €/m^2 after launching the ship. They also reported costs for coating repair work. Coating repair costs at dry-dock were estimated to 70 €/m^2. Costs reported for coating maintenance performed on offshore installations were significantly higher, 2500–3000 €/m^2.

Knudsen *et al.* (2012) and Knudsen (2010) evaluated different existing coating systems for structural components of wind turbines. Costs for application of protective coating systems on new structures, including grit blasting prior to painting were estimated as shown in Table 7.7. Costs for coating repair offshore were also briefly discussed. Maintenance of coating systems in areas with reasonable access on offshore oil and gas installations was estimated to be approximately 150 €/m^2. Coating maintenance costs for offshore turbines were expected to be significantly higher

Table 7.6 *Costs of painting offshore structures in a paint shop compared to costs for coating repair work.*

Coating application/ coating repair	*Costs*	*Comment*
Application in the paint shop (new constructions)	15–25 €/m^2	Costs depend on the setup and the paint system
Repair work onshore	75–250 €/m^2 5–10 x costs in the paint shop	At a construction site
Repair work offshore (with all job-related costs accumulated)	>1000 €/m^2 or >100 x costs in the paint shop which indicates costs >1500–2500 €/m^2	High costs for offshore is due to several factors including - Logistics of getting workers and materials to job site - Limited access to the structures

Data source: Mühlberg (2010).

Table 7.7 *Costs for painting of structural parts of offshore wind turbines based on tenders for coating of a Norwegian bridge (2008) and information based on input from Statoil and coating suppliers.*

Zone	*Grit blasting and coating application*	*Costs[a] [NOK/m²]*
Atmospheric	4-coat paint system	400
	3-coat paint system	320
	Duplex coating system TSZ with a sealer, and a 2-coat paint system	520
	TSA with a sealer	600
Splash	Reinforced polyester coating	560
Submerged	2-coat epoxy coating	280
Internal surface	Duplex coating system TSZ with a sealer and a 2-coat epoxy coating	480
	2-coat paint systems	240

Data sources: Knudsen *et al.* (2012) and Knudsen (2010).
[a]Cost figures converted from euro (€) to Norwegian krone (NOK), 1 € = 8 NOK.

compared to offshore oil and gas installations due to accessibility problems. Using long life maintenance free coatings such as duplex coating systems was therefore assumed to be beneficial from an LCC perspective.

7.5 Areas that need to be protected by coatings

To be competitive with onshore wind projects, larger and more reliable turbines and design optimizations are necessary for offshore wind parks in offshore environments (Henderson *et al.*, 2003; Breton and Moe, 2009). The amount of energy captured can be increased due to more wind. Transportation, installation, operational and maintenance costs can be reduced by enabling customers to run fewer, larger turbines, with fewer service visits. Furthermore, it is argued that the larger the wind turbine, the greener the electricity will be (Caduff *et al.*, 2012). Vestas has now developed V164-8.0 (Vestas, 2014), an offshore wind turbine with 8.0 MW capacity. A test rig of the V164-8.0 turbine will be installed in 2014 at the Danish national wind turbine test centre at Østerild (Wittrup, 2014; WindPower Directory, 2014). The 8.0 MW turbine is reported to be the world's largest capacity wind turbine at present. The turbine has a rotor diameter of 164 m, 80 m long blades, a tower height of 140 m and an overall height of 220 m.

The size of the wind turbine and the type of foundation used are affecting the structural dimensions of an offshore wind turbine. The height of the offshore wind turbine is increasing, caused by the increasing turbine capacity and the size of the turbine blades. The height of the SMZ of offshore installations depends on the geographical location and type of construction, i.e. whether we have a bottom fixed or a floating substructure (GL, 2012; DNV GL, 2013). For a floating construction in the North Sea, the SPZ will be from 4.1 m below the water line to 9.7 m above the water line

Table 7.8 *Dimensions of offshore wind turbines for two examples: Belwind Offshore Wind Farm and Hywind.*

Project	*Belwind I*	*Hywind*
Turbines	55 Vestas V90-3.0 MW	Siemens 2.3 MW
Rotor diameter	90 m	82.4 m
Hub height	72 m	65 m
Tower	55 m in height	46.5 m in height 5 m in diameter at base 2.4 m in diameter at top
Transition piece	25 m, 17 m above water line Diameter 4.3 m	28 m, 17 m above water line Diameter reduced from 8.3 to 6 m at water line Diameter 5 m at top
Foundation	50–72 m monopile, driven 35 m into seabed Diameter 5 m	Floating structure that reaches 100 m below water line Diameter 8.3 m

(Knudsen and Bjørgum, 2009). Thus, the main part of the transition piece is in the SPZ, and needs to be protected in accordance with the SPZ requirements. For offshore wind turbines with bottom-fixed monopile foundations, anodes for cathodic protection of submerged parts are usually attached to the transition piece. Usually, the monopile is not coated due to cost saving reasons and due to mechanical stress during installation when driving the monopile into the seabed. The hub heights for offshore wind turbines are typically 60–105 m (AWS Truewind, 2010) and are currently increasing to 120 m (European Energy Association, 2009) or more due to the increasing turbine size. Thus, there are large variations in the areas of the support structures to be protected by coatings. Detailed information about structural dimensions is usually not publically available, but in Table 7.8 dimensions of wind turbines in the Belwind Offshore Wind Farm (Lindoe Offshore Renewables Center, 2011) and for the floating Hywind turbine (Statoil, 2012) are shown as examples.

7.6 Life cycle cost analysis of different protective coating systems

In this section, LCC of different coating systems used to protect structural steel in the AZ of offshore wind turbines are analysed. The following costs are considered:

- Investment costs for applying coating systems in the paint shop
- Maintenance costs (coating repair in AZ)
    - Costs for surface cleaning and for applying coatings offshore
    - Access costs including costs for vessels and waiting for feasible weather window
    - Power production loss

Since the main focus is on comparing different coating systems and to find the most cost effective coating solution, cost elements that are not influenced by the choice of coating system (e.g. surface cleaning), or cost elements that are considered to be negligible, are not taken into account in the analysis.

All maintenance costs that incur in year N after installation are discounted with rate r to calculate the net present value (NPV). Thus, the NPV of the total costs C can be calculated as follows:

$$C = C_{\text{investment}} + C_{\text{maintenance}}\,(1 + r)^{-N}$$

The investment cost $C_{\text{investment}}$ is

$$C_{\text{investment}} = A_{\text{AZ}} \cdot c_{\text{AZ}} + A_{\text{SPZ}} \cdot c_{\text{SPZ}} + A_{\text{SMZ}} \cdot c_{\text{SMZ}}$$

where A is the area to be protected by coating and c the cost for applying the coating onshore in the paint shop (cost per m^2 coating; including surface pretreatment) for, respectively, the atmospheric zone (AZ), the splash zone (SPZ) and the submerged zone (SMZ).

The maintenance cost $C_{\text{maintenance}}$ is the sum of the costs for coating maintenance (C_{CM}, i.e. costs for applying the coating offshore), the costs for accessing the offshore wind turbine (C_{AC}) and the costs for production loss when carrying out maintenance (C_{PL}):

$$C_{\text{maintenance}} = C_{\text{CM}} + C_{\text{AC}} + C_{\text{PL}}$$

The costs for coating maintenance can be simply calculated by

$$C_{\text{CM}} = A_{\text{CM}} \cdot c_{\text{CM}}$$

where A_{CM} is the coating area to be maintained and c_{CM} is the cost for repairing offshore one m^2 of coating (this includes cost for coating, applying the coating, preparing surfaces, but excluding turbine access, waiting times and production loss). Assuming that three wind turbines can be maintained at the same time/day, access costs are approximated by

$$C_{\text{AC}} = \tfrac{1}{3}(t_{\text{w}} \cdot N_{\text{p}} \cdot c_{\text{p}} + t_{\text{v}} \cdot c_{\text{v}})$$

where t_{w} is the waiting time in hours for a suitable time window where it is possible to access the turbine, N_{p} is the number of personnel required for carrying out coating maintenance, c_{p} is the hourly rate of the personal, t_{v} is the number of hours a vessel is required and c_{v} is the hourly rate for the vessel. We assumed that travelling time to the wind farm is negligible compared to working time and waiting time. The production loss is calculated by

$$C_{\text{PL}} = t_{\text{PL}} \cdot P \cdot p_{\text{el}}$$

where t_{PL} is the number of hours with production loss, P is the average production in periods where maintenance is carried out and p_{el} is the electricity price.

A Monte Carlo approach is used to consider the uncertainty of several input parameters; for example, cost for coatings, cost for vessels and waiting time for weather. The analysis was carried out in MS Excel. The Excel add-in MCSim (Barreto and Howland, 2005) was used for performing Monte Carlo simulations.

Table 7.9 summarizes the general input parameters that are assumed to be equal for all coating systems and therewith for all LCCA cases. Some input parameters are changing dependent on the type of coating system. Especially, the total working time for applying the coating and the limits for temperature and humidity are changing from case to case. These parameters are summarised in Table 7.10. Coating

Table 7.9 *General input parameters (case independent).*

Parameter	*Value*	*Comment*
Interest rate	5–10%	
Atmospheric zone coating area (tower)	874 m^2	Dimensions from the Hywind turbine
Splash zone coating area (transition piece)	380 m^2	Dimensions from the Hywind turbine
Submerged zone coating area (submerged parts of sub-structure and foundation)	2,522 m^2	Dimensions from the Hywind turbine
Assumed maintenance coating area	100 m^2	Assumption: Maintenance can only be carried out in atmospheric zone (tower; see Tables 11–18 for details about coatings and maintenance costs in the atmospheric zone)
Investment cost for applying coating onshore (splash zone)	560 NOK/m^2	2-coat thick reinforced polyester coating
Investment cost for applying coating onshore (submerged zone)	280 NOK/m^2	2-coat epoxy coating
Wave limit access	2 m	Limiting significant wave height for typically used access vessels
Wind limit work	12 m/s	Limit for entering turbine and working on turbine
Number of personnel	2	
Personnel cost	1,000 NOK/hour	
Vessel time needed	16 hours (2-coat systems) 24 hours (3-coat systems)	Based on 2 hours travelling and 6 hours working per coating
Vessel cost	5,000–10,000 NOK/hour	
Number of hours with production loss	12 hours (2-coat systems) 18 hours (3-coat systems)	Based on 6 hours working per coating
Average power production during maintenance period	0.5 MW	Based on a 2.3 MW wind turbine and average production weather conditions when coating maintenance can be undertaken
Electricity price	1 NOK/kWh	Based on feed-in tariff in Germany

Table 7.10 *Individual maintenance related input parameters (case dependent).*

	Parameter	*Comment*
Investment	Cost for applying coating onshore (atmospheric zone)	Dependent on the type of coating.
Maintenance	Cost for applying coating offshore (atmospheric zone)	10 times the onshore cost. Assumption: Maintenance can only be carried out in atmospheric zone (tower)
	Working time total hours	Includes direct working time at turbine (6 hours per coating) and waiting time until coating is hardened
	Limit relative air humidity	Dependent on the type of coating
	Limit air temperature	Dependent on the type of coating
	Waiting time weather	Calculated based on the weather limitations (wind, wave, humidity, temperature) and the total working time, weather data from Gullfaks platform (yr. no. 2006–2012)
	Year of maintenance	Dependent on the type of coating

Table 7.11 *Case 1a: Maintenance free duplex coating systems.*

Description	*Parameter*	*Value*
Maintenance free coating systems, with duplex coating systems in the atmospheric zone	Investment cost for applying coating onshore (atmospheric zone)	520–600 NOK/m^2
	Cost for applying coating offshore (atmospheric zone)	Not applicable
	Working time total hours	Not applicable
	Limit relative air humidity	Not applicable
	Limit air temperature	Not applicable
	Waiting time weather	Not applicable
	Year of maintenance	Not applicable

specifications are taken from technical data sheets for representative protective coatings. Weather data from the Gullfaks oil field in the North Sea (yr. no. 2006–2012) are used to estimate waiting times for weather windows that are long enough to perform required maintenance operations.

Different cases are defined to compare the LCC of two typical maintenance free coating system (Cases 1a and b) with different 2-coat and 3-coat paint systems that may need maintenance after approximately 10 to 15 years in service (Cases 2–4). Tables 7.11 to 7.18 define the different cases that are used in the LCCA performed.

Table 7.12 *Case 1b: Maintenance free reinforced polyester 2-coat system.*

Description	*Parameter*	*Value*
Maintenance free 2-coat reinforced polyester system in the atmospheric zone	Cost for applying coating onshore (atmospheric zone)	500–620 NOK/m^2
	Cost for applying coating offshore (atmospheric zone)	Not applicable
	Working time total hours	Not applicable
	Limit relative air humidity	Not applicable
	Limit air temperature	Not applicable
	Waiting time weather	Not applicable
	Year of maintenance	Not applicable

Table 7.13 *Case 2a: 3-coat system, maintenance at 5°C.*

Description	*Parameter*	*Value*
3-coat system with maintenance need after 10–15 years (zinc epoxy primer, epoxy mastic, polyurethane) 3 jobs each 6 hours 3 turbines in parallel (weather waiting costs and vessel costs are divided by 3)	Cost for applying coating onshore (atmospheric zone)	320–400 NOK/m^2
	Cost for applying coating offshore (atmospheric zone)	3,200–4,000 NOK/m2
	Working time total hours	48 hours
	Limit relative air humidity	95%
	Limit air temperature	5°C
	Waiting time weather	130–210 hours
	Year of maintenance	10th–15th year

Table 7.14 *Case 2b: 3-coat system, maintenance at 10°C.*

Description	*Parameter*	*Value*
3-coat system with maintenance need after 10–15 years (zinc epoxy primer, epoxy mastic, polyurethane) 3 jobs each 6 hours 3 turbines in parallel (weather waiting costs and vessel costs are divided by 3)	Cost for applying coating onshore (atmospheric zone)	320–400 NOK/m^2
	Cost for applying coating offshore (atmospheric zone)	3,200–4,000 NOK/m^2
	Working time total hours	42 hours
	Limit relative air humidity	95%
	Limit air temperature	10°C
	Waiting time weather	150–200 hours
	Year of maintenance	10th–15th year

Table 7.15 *Case 3: 2-coat system, epoxy mastic.*

Description	*Parameter*	*Value*
2-coat system with maintenance need after 10–15 years (epoxy mastic) 2 jobs each 6 hours 3 turbines in parallel (weather waiting costs and vessel costs are divided by 3)	Cost for applying coating onshore (atmospheric zone)	280 NOK/m^2
	Cost for applying coating offshore (atmospheric zone)	2,800 NOK/m2
	Working time total hours	36 hours
	Limit relative air humidity	85%
	Limit air temperature	10°C
	Waiting time weather	480–490 hours
	Year of maintenance	10th–15th year

Table 7.16 *Case 4a: 2-coat system, epoxy mastic winter grade, maintenance at 0°C.*

Description	*Parameter*	*Value*
2-coat system with maintenance need after 10–15 years (epoxy mastic winter grade) 2 jobs each 6 hours 3 turbines in parallel (weather waiting costs and vessel costs are divided by 3)	Cost for applying coating onshore (atmospheric zone)	280 NOK/m^2
	Cost for applying coating offshore (atmospheric zone)	2,800 NOK/m^2
	Working time total hours	38 hours
	Limit relative air humidity	85%
	Limit air temperature	0°C
	Waiting time weather	120–270 hours
	Year of maintenance	10th–15th year

Table 7.17 *Case 4b: 2-coat system, epoxy mastic winter grade, maintenance at 5°C.*

Description	*Parameter*	*Value*
2-coat system with maintenance need after 10–15 years (epoxy mastic winter grade) 2 jobs each 6 hours 3 turbines in parallel (weather waiting costs and vessel costs are divided by 3)	Cost for applying coating onshore (atmospheric zone)	280 NOK/m^2
	Cost for applying coating offshore (atmospheric zone)	2,800 NOK/m^2
	Working time total hours	30 hours
	Limit relative air humidity	85%
	Limit air temperature	5°C
	Waiting time weather	100–170 hours
	Year of maintenance	10th–15th year

Table 7.18 *Case 4c: 2-coat system, epoxy mastic winter grade, maintenance at 10°C.*

Description	*Parameter*	*Value*
2-coat system with maintenance need after 10–15 years (epoxy mastic winter grade) 2 jobs each 6 hours 3 turbines in parallel (weather waiting costs and vessel costs are divided by 3)	Cost for applying coating onshore (atmospheric zone)	280 NOK/m
	Cost for applying coating offshore (atmospheric zone)	2,800 NOK/m^2
	Working time total hours	24 hours
	Limit relative air humidity	85%
	Limit air temperature	10°C
	Waiting time weather	230–280 hours
	Year of maintenance	10th–15th year

Parameters that are given as a range in Tables 7.9 to 7.18 are modelled as uniformly distributed between the minimum and maximum values given in the tables. All costs are given in NOK (Norwegian krone; 8 NOK $\approx$ 1 €).

7.6.1 Analysis results

The magnitude and the variation of the NPV of costs for the different cases are illustrated by the distributions in Figure 7.2. The reason for the NPV of the costs being presented as a discrete probability distribution is the uncertainty related to some of the LCCA input parameters.

Although NORSOK M-501 (2012) does not recommend 2-coat systems for corrosion protection in the AZ of offshore installations, we have included such a coating system in the LCCA for comparison. Despite that the coating maintenance are needed, 2-coat systems based on epoxy mastic (Cases 3 and 4) and 3-coat systems including a polyurethane topcoat (Case 2) show LCC similar to the maintenance free coating system. In fact, the coatings that can be used in low temperatures (Cases 4a, 4b and 4c) perform best from a LCC perspective even though the differences are small. Apart from the epoxy mastic winter grade 2-coat systems, all other coatings need temperatures above 10°C over a longer period for the maintenance work, and these weather conditions are rare in the North Sea. This limitation leads therefore to higher maintenance costs due to the long waiting times for feasible weather windows.

Note that the cost differences between the cases are not significant when taking into account the uncertainty of the results. This is illustrated by the considerable overlap of the probability distributions in Figure 7.2.

7.7 Conclusions

Protective coatings for offshore wind turbines have been evaluated with respect to expected technical performance and, when available, based on experience from service. Based on the available information reviewed, experiences from maintenance of offshore wind turbine coatings generally seem to be insignificant, and design for a long lifetime is the main strategy. Experiences from the offshore oil and gas industry

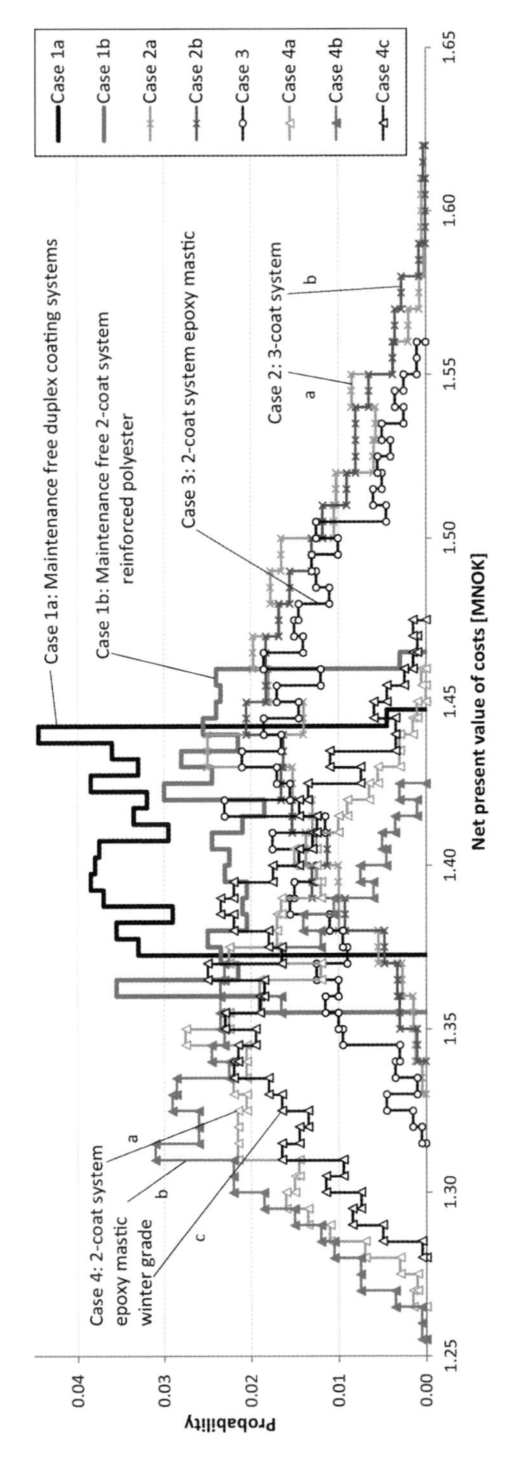

Figure 7.2 *Net present value of costs [MNOK] per turbine for different coating systems.*

Table 7.19 *Mean value, standard deviation and min/max values of the NPV for all cases [MNOK].*

Case	*Mean*	*Standard deviation*	*Min*	*Max*
1a	1.409	0.021	1.374	1.443
1b	1.408	0.031	1.356	1.461
2a	1.458	0.049	1.336	1.610
2b	1.459	0.051	1.340	1.616
3	1.433	0.051	1.320	1.558
4a	1.351	0.039	1.266	1.455
4b	1.336	0.034	1.259	1.425
4c	1.367	0.041	1.280	1.473

show that coating systems exist that satisfy the demanded 25 years of lifetime for offshore wind turbines. However, high investment costs may be a challenge for large offshore wind farms and may therefore prevent the use of these coating systems.

Based on cost evaluations for coating application, a LCCA has been carried out for different corrosion protective coating systems for structures in the AZ on offshore wind turbines. Weather conditions used for the analysis are typical for the North Sea. A Monte Carlo approach has been used to take into account uncertainties in the input parameters and to illustrate uncertainties related to the analysis results.

The results from the analysis show that, apart from 2-coat epoxy mastic winter grade systems which can be applied between 0°C and 10°C, the LCC for maintenance free coatings are generally lower than for typical 3-coat and 2-coat systems that need maintenance after 10 to 15 years in service. The maintenance free coating systems evaluated here are so called duplex coatings consisting of TSZ and a 2-coat reinforced polyester coating system normally used in the SPZ. Although 2-coat epoxy systems are not recommended for the AZ on offshore installations, such coating systems are included here for comparison.

It is important to note that the aforementioned conclusions are mainly based on the mean value of the LCC. When taking into account the uncertainties obtained by the Monte Carlo approach used in the LCCA, it must be concluded that general recommendations are difficult to make. The advice is therefore to carry out a more detailed analysis when decisions about the choice of coatings for a specific installation are made. For some of the input parameters that are considered in the present work as quite uncertain, better estimates can be created when analysing a real application. Then, the uncertainty of the results can be reduced, and it will be possible to make decisions based on more significant results.

Acknowledgement

This study has been performed in connection with the Norwegian Research Centre for Offshore Wind Technology—NOWITECH. NOWITECH is a research programme which focuses on offshore wind energy in "deep-sea" areas (+30 m). Both bottom-fixed and floating wind turbines are addressed, and protective coatings for offshore wind turbines are one of the research topics. The financial support by the Research Council of Norway and the NOWITECH partners is gratefully acknowledged.

References

AWS Truewind (2010) *Offshore wind technology overview, nyserda pon 995, task order no. 2, agreement no. 9998*. AWS Truewind for the Long Island New York City Offshore Wind Collaborative. http://www.linycoffshorewind.com/PDF/AWS%20Truewind%20Offshore%20Wind%20Technology%20Final%20Report.pdf. Accessed 7/7/14.

Axelsen, S.B., A. Sjaastad, O.Ø. Knuden and R. Johnsen (2010) Protective coatings offshore: introducing a risk based maintenance management system-part 3: a case study. *Corrosion* **66**, 015004-1–015004-10.

Barreto, H. and F.M. Howland (2005) *Introductory econometrics. Excel add-in: Monte Carlo simulation*. http://www3.wabash.edu/econometrics/EconometricsBook/Basic%20Tools/ExcelAddIns/MCSim.htm. Accessed 7/7/14.

Breton, S.-P. and G. Moe (2009) Status, plans and technologies for offshore wind turbines in Europe and North America. *Renewable Energy* **34**, 646–654.

Caduff, M., M.A.J. Huijbregts, H.J. Althaus, A. Koehler and S. Hellweg (2012) Wind power electricity: the bigger the turbine, the greener the electricity? *Environ. Sci. Technol.* **46**, 4725–4733.

Corrpro Companies Inc. (2001) *Cost effective alternative Methods for Steel Bridge Paint System Maintenance, Contract No. DTFH61-97-C-00026, Report IX: Field Metallizing Highway Bridges*. Written for the Federal Highway Administration, Washington. Lydia Frenzel Conference Series, Advisory Council-Ac10014.

DNV GL (2012) *Guideline for the certification of offshore wind turbines, DNV GL*. http://www.gl-group.com/en/certification/renewables/26664.php. Unable to access on 7/7/14.

DNV GL (2013) *Guidelines and technical notes for the certification of wind turbines, wind farms and other renewables*. Available from: http://www.gl-group.com/en/certification/renewables/CertificationGuidelines.php.

DNV-OS-F101 (2012) *Submarine pipeline systems*, Det Norske Veritas AS, DNV Offshore Standard. August. Available from http://www.dnv.com

DNV-OS-J101 (2014) *Design of offshore wind turbine structure*. Det Norske Veritas AS, DNV Offshore Standard. May 2014. Available from http://www.dnv.com

Dong Energy (2012) *Personal communication with Kringelum, J.V.*

European Energy Association (2009) *Europe's onshore and offshore energy potential—an assessment of environmental and economic constraints*, EEA Technical Report, No. 6. European Environment Agency, Copenhagen, Denmark.

GL Renewables Certification (2012) *Rules and Guidelines, IV: Industrial Services, 2: Guideline for the Certification of Offshore Wind Turbines. Chapter 3: Requirements for Manufacturers, Quality Management, Materials, Production and Corrosion Protection*. Germanische Lloyd Wind Energie, Hamburg, Germany.

Henderson, A.R., C. Morgan, B. Smith, H.C. Sørensen, R.J. Barthelmie and B. Boesmans (2003) Offshore energy in Europe—a review of the state-of the art. *Wind Energy* **6**(1), 35–52.

Herion, S., T. Faber and J. Hrabowski (2009) Extension of service life and considerations on corrosion problems of offshore wind energy converters. In *Proceedings of the Nineteenth International Offshore and Polar Engineering Conference*. Osaka, Japan.

ISO12944-4 (1998) Paints and varnishes—corrosion protection of steel structures by protective paint systems, in Part 4—types of surface and surface preparation. 1998-05.

ISO12944-5 (2007) Paints and varnishes—corrosion protection of steel structures by protective paint systems, in Part 5—protective Paint Systems. 2007-09-15.

ISO 20340 (2009) Paints and varnishes—performance requirements for protective paint systems for offshore and related structures. 2009-04-01.

Kattan, R. (2011) *Introduction to Safinah Ltd. Managing your coating risk – minimising your costs*. London's Thursday Lecture Series, Lecture 64; Bramar Adjusting: http://www.braemaradjusting.com/news/newsevents/thursday-lectures/

Klinge, R. (1999) Protection of Norwegian steel bridges against corrosion. *Stahlbau* **68**(5), 382–391.

Klinge, R. (2009) Altered specifications for the protection of Norwegian steel bridges and offshore structures against corrosion. *Steel Construction* **2**(2), 109–118.

Knudsen, O.Ø. (2010) *Evaluation of existing coatings for corrosion protection of structural components: Nowitech WP3*. 2010-12-17, SINTEF Report F17552, Trondheim, Norway.

Knudsen, O.Ø. and A. Bjørgum (2009) Corrosion protection of offshore wind turbines—long life protective coatings. In *EWEC 2009 Proceedings*. Marseille, France, March.

Knudsen, O.Ø., A. Bjørgum and L.T. Døssland (2012) Low maintenance coating systems for constructions with long lifetime, in *Corrosion 2012*. NACE International, Salt Lake City, Utah.

Knudsen, O.Ø., T. Røssland and T. Rogne (2004). Rapid degradation of painted TSA, Paper No 04023. In *Corrosion 2004,* NACE. NACE , New Orleans, Louisiana.

Lindoe Offshore Renewables Center (LORC) (2011) *Belwind 1 offshore wind farm*. Available from: http://www.linycoffshorewind.com/PDF/AWS%20Truewind%20Offshore%20Wind%20Technology%20Final%20Report.pdf, Accessed 7/7/14; http://www.lorc.dk/offshore-wind-farms-map/belwind-1. Accessed 7/7/14.

Momber, A. (2011) Corrosion and corrosion protection of support structures for offshore wind energy devices (OWEA). *Materials and Corrosion* **62**(5). 391–404.

Mühlberg, K. (2010) Corrosion protection of offshore windturbines—a challenge for the steel builder applicator. *Journal of Protective Coatings & Linings* **27**(3). 20–33.

NORSOK M-001, N. (2004) Materials selection. Rev. 4. August. Standards Norway, Lysaker, Norway.

NORSOK M-501 (2012) Surface preparation and protective coating. Rev. 6, February. Standards Norway, Lysaker, Norway.

NOWITECH (2014) *The Norwegian research centre of offshore wind technology*. http://www.sintef.no/Projectweb/Nowitech/. Accessed 7/7/14.

Petersen, P., B. Bendix and T. Feld (2005) *Design basis for offshore wind structures*. Copenhagen offshore wind. http://wind.nrel.gov/public/SeaCon/Proceedings/Copenhagen.Offshore.Wind.2005/documents/papers/Design_basis/P._Petersen_Design_basis_for_offshore_wind_structures.pdf. Accessed 07/07/14.

Quarton, D.C. (2005) *An international design standard for offshore wind turbines: IEC 61400-3*. Garrad Hassan and Partners Ltd, St Vincent's Works, Copenhagen Offshore Wind. http://wind.nrel.gov/public/SeaCon/Proceedings/Copenhagen.Offshore.Wind.2005/documents/papers/Design_basis/D.Quarton_An_international_design_standard_for_offshore.pdf. Accessed 07/07/14.

Safinah Ltd (2011) *Introduction to Safinah Ltd. Managing your coating risk—minimising your costs*. Available from: www.braemarsteege.com/lecturenotes/lecture64.pdf. Unable to access 7/7/14.

Statoil (2012) *Hywind: putting wind power to the test*. http://www.statoil.com/en/TechnologyInnovation/NewEnergy/RenewablePowerProduction/Onshore/Pages/Karmoy.aspx. Could not be accessed on 7/7/14.

Van der Mijle Meijer, H. (2009) Corrosion in offshore wind energy – a major issue. In *the Offshore Wind Power Conference Essential Innovations*. http://www.we-at-sea.org/wp-content/uploads/2009/02/3-Harald-vd-Mijle-Meijer.pdf. February 12–13, 2009. Den Helder, The Netherlands. Accessed 07/07/14.

Vests (2014) *Vestas V164-8.0 MW® at a glance*. http://vestas.com/en/products_and_services/offshore#!v164-development. Accessed 7/7/14.

WindPower Directory (2014) *The 10 biggest turbines in the world.* http://www.windpowermonthly.com/10-biggest-turbines. Accessed 7/7/14.

Wittrup, S. (2014) Power from Vestas' giant turbine (in Danish). *Ingeniøren*, 28 January.

yr. no. (2006–2012) Detailed weather statistics for Gullfaks C observation site from the Norwegian Meteorological Institute and the Norwegian Broadcasting Corporation (NRK), Available from: http://www.yr.no/place/Norway/Hav/Gullfaks_C_observation_site/detailed_statistics.html.

Index